GUIDE TO
WEATHER
FORECASTING

STORM DUNLOP

First published in 2008 by Philip's,
a division of Octopus Publishing Group Limited
(www.octopusbooks.co.uk)
Endeavour House, 189 Shaftesbury Avenue,
London WC2H 8JY
An Hachette UK Company (www.hachette.co.uk)

Second edition published 2013

Copyright © 2008, 2013 Storm Dunlop

ISBN 978–1–84907–303–5

Printed in Thailand

Details of other Philip's titles and services
can be found on our website at:
www.philips-maps.co.uk

FRONT COVER : *tl* Multicell thunderstorm, *Corbis*;
tr January temperature for North America,
Philip's; *c* Hand-held anemometer, *Richard Paul
Russell Ltd.*; *bl* Synoptic chart, *Philip's*;
br Cirrus intortus, *Corbis*
BACK COVER : *bl* Meteosat, *EUMETSAT*;
br diagram of alternating lobes of cold and
warm air, *Philip's*

Contents

THE ORIGINS OF WEATHER

THE ATMOSPHERE AND THE GLOBAL CIRCULATION

It may seem a little strange to begin talking about weather forecasting by describing conditions over the whole Earth, but in fact it makes sense. Forecasting the weather requires a knowledge of the situation over a large area 'upwind' (so to speak) of the area in which you are interested. To forecast just one day ahead, professional weather forecasters in Europe, for example, need to know what is happening right across the Atlantic. Similarly, forecasters on the West Coast of North America require details of the situation across the Pacific as far as Japan. In preparing forecasts for three days ahead, forecasters need detailed information about conditions across the whole Earth, including data from the southern hemisphere and Antarctica. An understanding of the basic mechanisms driving the weather is extremely helpful for predicting what is going to happen on even a local scale.

The structure of the atmosphere

Most weather phenomena, including the majority of clouds, occur in the lowest layer of the atmosphere, the troposphere. This is extremely thin compared with the size of the Earth, which has an equatorial diameter of 12,756 km (7,926 mi) and 12,714 km (7,900 mi) measured across the poles. Yet the troposphere extends to about 18–20 km (11–12 mi) at the most (in the equatorial regions) and to just about 7 km (4.3 mi) at the poles. The level of the top of the troposphere, called the tropopause, is defined by a change in the way the

▶ A typical temperature profile of the atmosphere. The height of the tropopause varies from about 18–20 km (11–12 mi) at the equator to about 7 km (4.3 mi) at the poles. The mesopause varies in altitude from summer to winter.

▼ The anvils of these cumulonimbus clouds lie at the tropopause, here, near the equatorial region, about 15 km (9 mi) high.

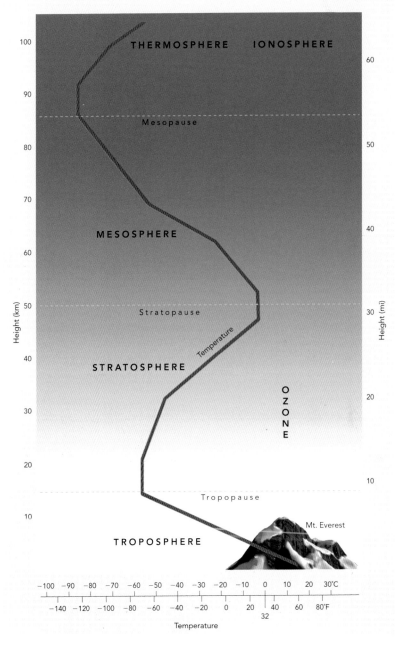

temperature behaves with increasing altitude. Between the surface and the tropopause the temperature generally decreases with height – albeit often in an irregular manner. The changes in temperature with height are extremely important for the formation of clouds, as we shall see later (p.24).

At the tropopause, the decline ceases, and the temperature tends to remain constant in the lowest region of the next layer, the stratosphere (p.5). It then starts to increase with height, reaching a maximum at an altitude of about 50 km (31 mi). This heating in the stratosphere is the result of the absorption of ultraviolet radiation from the Sun by molecules of ozone, O_3, whose greatest concentration occurs at about 20–25 km (12–15.5 mi). The destruction of this ozone by man-made chemicals has led to the formation of the seasonal 'ozone holes' over the Antarctic and Arctic regions. There are few clouds in the stratosphere, although sometimes there are ice-crystal clouds in the lowermost region, including, on rare occasions, beautiful nacreous clouds (pp.51–2).

At the top of the stratosphere, at the stratopause, which lies at an altitude of approximately 50 km (31 mi), the temperature again begins to decrease with height within the layer known as the mesosphere. The very lowest temperatures in the atmosphere (–163 to –100°C (–260 to –148°F)) are found at the top of the mesosphere, at the mesopause, which is generally at an altitude of about 86 km (53 mi) (or roughly 100 km (62 mi) over the polar regions in summer). Conditions in the mesosphere have no direct effect on the weather down at the surface, but the highest clouds in the atmosphere, noctilucent clouds (pp.52–3), occur just below the mesopause and are sometimes visible from high latitudes, in summer, in the middle of the night.

Above the mesosphere lies the outermost layer of the atmosphere, the thermosphere. The uppermost region of the mesosphere and lowest part of the thermosphere (between 60 and 1,000 km (40 and 620 mi), approximately) is also known as the ionosphere. Although this region is significant for communications because of its effects on radio waves, and is also the site of the aurorae, it, and the thermosphere in general, have little direct effect on the weather at surface level.

The ozone holes

By absorbing ultraviolet radiation, the ozone in the stratosphere shields the ground from its harmful effects. Ozone is broken down by persistent man-made chemicals, particularly by the substances known as chlorofluorocarbons (CFCs). Release of these has led to a decline in stratospheric ozone by several per cent worldwide, with major depletions in the polar regions. There, rapid breakdown occurs on the surfaces of the particles in polar stratospheric clouds (including nacreous clouds, pp.51–2), particularly when sunlight returns to the polar regions after the winter night.

The ozone hole is especially large over Antarctica, where the particularly strong westerly winds (the polar vortex) tend to isolate the region from the rest of the global circulation. Following the banning of CFCs and other chemicals by international treaty in 1987, there are some indications that the ozone holes have begun to decline, but full recovery is likely to take decades.

▶ *Pale silvery-blue noctilucent clouds are visible high above the orange-coloured troposphere and the dark, low-level clouds. The atmosphere extends even farther into space and its effect may be seen by the slight blurring of the lower cusp of the waning Moon.*

Composition of the atmosphere

Up to a height of about 85 km (53 mi), i.e., roughly to the top of the mesosphere – the composition of the air is largely constant. It consists of:

nitrogen	78.09%
oxygen	20.95%
argon	0.94%
carbon dioxide	0.03%

with trace amounts of neon, helium, methane, krypton, hydrogen, nitrous oxide and xenon. Water vapour ranges between 0 and 4 per cent, and the humidity of the air has important consequences that will be described later.

The layer in which gases are thoroughly mixed is known as the homosphere, but above about 85 km (53 mi) (in the region known as the heterosphere) the gases separate by mass, with the lightest gases (hydrogen and helium) in the outermost region, from which they escape into space.

Motion in the atmosphere

The motion of the atmosphere and the changes of the weather are ultimately driven by the heat of the Sun, and by the contrast in temperature between areas that are subjected to the full amount of heating and regions that receive less or no heating. Heat is carried from the hottest areas (generally in the tropics) towards the poles by winds and also by oceanic currents. Although differential heating is the ultimate cause of atmospheric motions, it acts by causing differences in pressure, which drive the planetary winds. The winds, in turn, drive the ocean currents, although the exact directions and strengths of the latter are, of course, also determined by the shape, size and depth of the ocean basins.

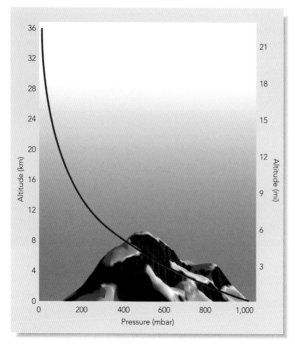

◀ An idealized diagram of the way in which pressure declines with height.

Pressure

Atmospheric pressure is the single most important factor in determining current and future weather. The air pressure at any particular point is related to the weight of the column of air above that point. Pressure therefore declines with height, but air is compressible, so its density is greatest at the surface and decreases with altitude. Pressure and density are also related to temperature: heating air causes the individual molecules of gas to move at a higher velocity. In a closed container this would cause the pressure to rise, but in the atmosphere, it causes the air to expand and its density to decrease. Cooling, naturally, causes the opposite effects: contraction of the air and an increase in density.

The differences in heating that exist across the globe therefore create differences in surface pressure. The distribution of high- and low- pressure regions will, of course change with the seasons, and the diagrams on the opposite page show the

Measuring pressure

Pressure is measured by a barometer (see p.134), and was initially quoted as inches of mercury (inHg). Such values are sometimes given in weather forecasts (primarily in the United States). More commonly, a unit known as the millibar (mb) is used – nominally one-thousandth of a bar (the approximate value for sea level). In fact, the average sea-level pressure has been defined as 1013.2 mb. Technically, pressure should be measured in pascals (Pa) or kilopascals (kPA). For convenience, however, and to prevent confusion, many meteorologists use another unit, the hectopascal (hPA), where 1 hectopascal (1 hPa) is identical to 1 millibar (1 mb).

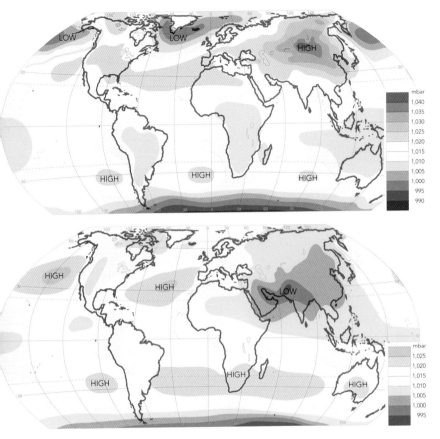

▲ *Mean sea-level pressure for January (top), showing the Siberian High and Icelandic and Aleutian Lows; and for July (bottom), where the south-western Asian low creates the Asian monsoon but pressures are never very high over the Southern Ocean.*

average distribution of surface pressure for January (top) and July (bottom). Here the areas of the globe with high and low pressures are shown in different colours, but on the charts used for forecasting (discussed later, p.149) the distribution of pressure is shown by isobars – lines joining points of equal pressure.

The most striking variations between the two patterns are the dramatic change over Siberia between high pressure in winter (the 'Siberian High') to low pressure in summer, and the low-pressure regions that arise in winter over the northern Atlantic and Pacific Oceans (the 'Icelandic Low' and the 'Aleutian Low', respectively). We shall return to both of these points shortly.

Differences in pressure set up a pressure gradient, causing air to flow from high-pressure regions to areas of low pressure. The resulting winds show complex changes with the seasons and from year to year.

9

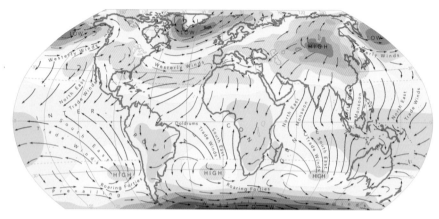

THE GLOBAL CIRCULATION

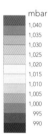

Rather than there being one large circulation cell in each hemisphere – as was originally believed – with heated air rising over the equatorial regions and descending over the cold poles, there are actually three circulation cells in each hemisphere.

A glance at any map showing the pattern of global winds such as here clearly reveals two of the three main systems in each hemisphere. Converging on the equatorial region (where winds are generally light or absent, in what are known as the Doldrums) are the trade winds: the north-easterly trades in the northern hemisphere and the south-easterly trades in the southern. Farther towards the poles, at temperate latitudes, there are the dominant westerlies, which are particularly strong in the southern hemisphere where there is little land to impede their progress, giving rise to the well-known Roaring Forties and the unofficially named Fearsome Fifties and Screaming Sixties. Even closer to the poles (and difficult to see on this particular map projection) there are bands of easterly winds blowing out from the poles towards temperate latitudes. The pattern of three systems is particularly well marked in the southern hemisphere, but in the north it is affected by the existence of large land masses, which tend to disrupt the pattern, especially in the winter.

The circulation cells

Air, heated at the equator, rises (creating a low-pressure zone at the surface) and then flows away to the north and south. It descends towards the surface at latitudes of approximately 30°N and 30°S. Here there are quasi-permanent regions of high surface pressure, known as the subtropical anticyclones or subtropical highs. From these

mbar
1,040
1,035
1,030
1,025
1,020
1,015
1,010
1,005
1,000
995
990

▲ *Mean sea-level pressure and winds for January.*

▼ *During the southern winter, Australia creates the only southern continental high.*

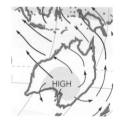

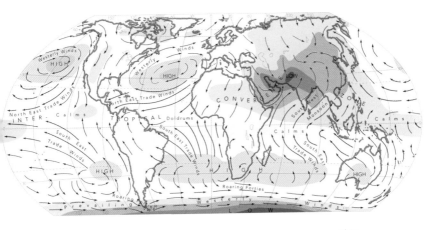

high-pressure regions, some of the air flows back towards the equator at low levels to complete the circulation cell, but some of the warm air flows out (also at the surface) towards the poles over the middle latitudes in each hemisphere.

At the same time, air that has become cooled over the polar regions becomes denser and flows from the poles towards the middle latitudes. It encounters the air flowing towards the poles along two atmospheric boundaries, one in each hemisphere and known as the Polar Fronts, where most of the weather systems that affect the temperate zones are generated. Along certain sections of the two Polar Fronts warm air rises over the colder air. The flow then divides, with some air flowing towards the poles and some returning towards the subtropical highs. There are therefore three main circulation cells within each hemisphere with some interchange of air between them.

The tropopause, the boundary that lies between the troposphere and the stratosphere, is not an unbroken surface.

mbar
1,025
1,020
1,015
1,010
1,005
1,000
995

▲ *Mean sea-level pressure and winds for July.*

▼ *Two high-pressure regions dominate the weather during the northern summer.*

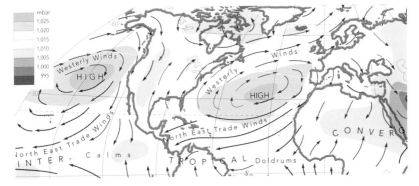

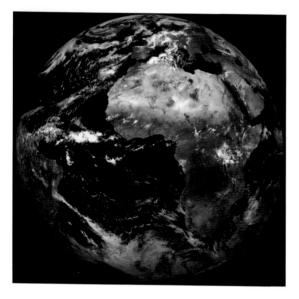

◀ In this Meteosat image, the northern Polar Front is the line of clouds running west from the Bay of Biscay. The subtropical highs over the Sahara and Arabia, and over southern Africa have largely clear skies. The cloud clusters over the equatorial Atlantic show the location of the Intertropical Convergence Zone where the north-easterly and south-easterly trade winds converge.

There may be distinct breaks and relatively abrupt changes of level, in particular near the location of the subtropical highs and close to the Polar Fronts. On occasions the tropopause may even be duplicated, with one layer lying at a greater altitude than the other. These breaks allow a certain interchange of air between the troposphere and the stratosphere, and the jet streams (which will be described shortly) often lie close to them.

WINDS AND THE CORIOLIS EFFECT

The direction of the wind is always described by the direction from which the air is moving, so the winds that tend to dominate the temperate regions of the Earth, blowing from west to east, are known as the westerlies. The reason for their dominance is that their direction is determined by the rotation of the Earth. The causes are worth describing in a little detail, because the effects are found even at local scales and have a bearing on current and forthcoming weather.

Because of the Earth's rotation, the air in the general circulation just described, does not flow directly north or south, but at an angle to the parallels of latitude. Consider a parcel of air at the equator (a, opposite). (Small volumes of air are often referred to as 'parcels' or 'packets' by meteorologists.) It is being carried eastwards at a considerable rate by the rotation of the Earth (more than 40,000 km (24,850 mi) in 24 hours), whereas at the poles any parcel of air is not moving at

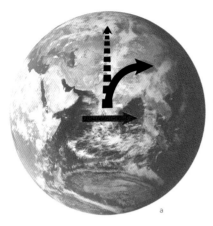

a

b

▲ The Coriolis effect: A parcel of air at the equator that moves north (a) is deflected to the east by the Earth's rotation, whereas a parcel moving towards the equator (b) is deflected to the west.

all, but is instead rotating once in 24 hours. The speed of the Earth's surface motion eastwards decreases from the equator to the poles.

If a parcel of air at the equator starts to move towards the pole, it tends to retain its original, rapid, eastward motion, but the surface below it is moving eastwards more slowly. This means that in the northern hemisphere, the parcel of air turns towards the right, relative to the surface below. In the southern hemisphere it turns to the left. This effect (known as the Coriolis effect) also occurs with air moving from the poles towards the equator (b, above). In this case, the air is moving more slowly than the surface it passes over, so it swings towards the right in the northern hemisphere, and to the left in the southern. The effect is weakest at the equator, and strongest at the poles, but is also dependent on the strength of the wind, being greater the faster the air is moving. This last point will be returned to later, because it plays an important role in controlling the flow of air in weather systems and is also relevant when determining local weather.

The trade winds, the westerlies and the polar easterlies

In the northern hemisphere, because of the Coriolis effect, the warm air that rises at the equator and then flows north is deflected towards the east – it becomes a generally south-westerly airflow at altitude – as it moves towards the sub-tropical anticyclones. The air from those anticyclones that moves south at the surface is similarly affected, becoming the generally north-easterly trade winds, blowing towards the equatorial zone. In the southern hemisphere, the effect produces the south-easterly trades.

The surface air flowing from the subtropical anticyclones towards the poles is similarly affected, and in both hemi-

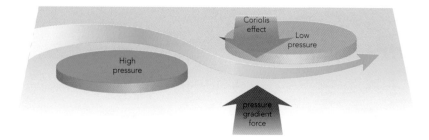

spheres the flow is deflected towards the east, giving rise to westerly winds over the middle latitudes. These tend to dominate the weather over the temperate zones where (particularly in the north) much of the Earth's population lives. In a similar manner, the cold air flowing from the poles is deflected to give rise to cold polar easterlies. These are always present in the southern hemisphere around the edge of Antarctica in the cold air flowing from the interior of the continent, but are less prominent in the north, where the pattern is disrupted in the winter by cold air flowing from Siberia and Arctic Canada.

▲ *The circulation around high- and low-pressure areas in the absence of friction with the surface (i.e., at a moderate altitude in the atmosphere).*

Circulation around highs and lows

As the maps show, in the northern hemisphere air flowing from one of the subtropical anticyclones actually circulates clockwise around the centre of the high (p.10). By contrast, air flowing towards a low-pressure region in the northern hemisphere circulates in an anticlockwise direction. Where this inflowing air forms a closed circulation, the low-pressure centre is known technically as a cyclone or, more commonly, as a low or depression. The direction may be remembered by what is called the Buys-Ballot Law (see box).

Contrary to what one might expect, air does not flow directly from high pressure to low pressure. As a parcel of air begins to move under the influence of the pressure gradient, the Coriolis force causes it to swing to the right in the northern hemisphere and to the left in the southern hemisphere. The deflection continues until the two forces (pressure gradient and Coriolis force) are in balance, exactly opposing one another, and the parcel of air circles the low pressure centre, following the curvature of the isobars. This situation does actually occur quite frequently in the atmosphere, away from the effects of friction with the ground. (This freely flowing air along the isobars is known technically as the

Where is the low? The Buys-Ballot Law

People often have difficulty in remembering which way the wind circulates around low-pressure centres. The simplest way is to use the Buys-Ballot Law, first popularized by the Dutch meteorologist, Hendrik Buys-Ballot: 'With your back to the wind, low pressure is on your left.' (That is in the northern hemisphere; in the southern hemisphere, the low pressure will be on your right.)

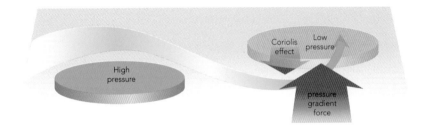

▲ The circulation around high- and low-pressure areas when the effect of friction with the surface is taken into account. The exact amount of convergence or divergence depends on the type of surface and the corresponding amount of friction.

geostrophic wind.) In the lowest layer of the atmosphere (known as the boundary layer) there is friction with the ground and associated turbulence. This reduces the speed of the wind and, because the two are linked, the strength of the Coriolis force. The moving parcel of air is not deflected as strongly towards the right (in the northern hemisphere), with the result that the surface wind spirals in towards the low-pressure centre, and spirals out from a high-pressure area.

The depth of the boundary layer is roughly 500 m (1,600 ft) over the sea and 1,500 m (5,000 ft) over the land. The practical result is that above those heights the wind is flowing in a slightly different direction from the surface wind. The difference amounts to about 10–15° over the sea and some 40–50° over the land. So the Buys-Ballot Law needs to be modified slightly, because, over the sea, the low-pressure centre will be about 10–15° farther forward than the law states, and over the land, perhaps as much as 40–50° farther forward. The wind is said to have veered by that amount. It is this difference in direction that often leads to confusion, because clouds will often be moving in a different direction to the surface wind.

Circulations in highs and lows are accompanied by vertical motions. The heated, rising air in the equatorial regions creates a low-pressure trough: a warm low. In the centre of a depression air converges, but cannot accumulate indefinitely, so is forced to rise, thus forming a cold low. There is a corresponding pair of high-pressure areas. Over the subtropical anticyclones the air is descending throughout the depth of the troposphere and creates a dry, warm high at the surface. In a cold high, air is descending and being compressed, which tends to warm it. However, the ground itself loses heat in the form of infrared radiation to space and cools the air immediately in contact with it. Because air conducts heat poorly,

Veering and backing

A wind is said to have veered, when the direction from which it is blowing has altered in a clockwise direction, i.e., towards the right. A change in the opposite direction (anticlockwise) is known as backing. Note that most meteorologists apply this usage to the winds in both hemispheres, but that a few American writers reverse the directions when describing winds in the southern hemisphere.

the layer of cold air is not heated by the slightly warmer air above. A pool of dense, cold air gradually accumulates and spreads out, so an even larger area starts to be cooled by contact with the ground.

Although the maps show that the circulation around the semi-permanent subtropical anticyclones does generally conform to the description just given, the circulation around the low-pressure centres is not as distinct as it appears. The two main winter low-pressure regions (the Icelandic Low and the Aleutian Low) appear on these averaged maps because true cyclones (depressions) with closed circulations frequently pass across them.

The monsoons
In the northern hemisphere, the most striking change from winter to summer is the switch from a cold high over Siberia in winter to a warm low in summer. This produces the two monsoons – the word derives from the Arabic word for 'season' – with a distinct change in wind direction. In winter the cold air flows out from the centre of the continent, giving rise to the north-east monsoon over the Indian Ocean and the north-west monsoon over other areas of south-east Asia. In summer, the winds are more or less reversed, with air flowing into the centre of Asia, bringing heavy rains from the tropical oceans. The arrival of the summer monsoon usually follows a period of extreme heat in early summer and despite the flooding that often accompanies it, is normally extremely welcome. Failure of the summer monsoon has major consequences for agriculture in India and south-east Asia generally, leading to droughts, crop shortages and even famine.

Air masses
When air remains stationary for a period over one particular part of the Earth, it tends to take two specific characteristics depending on its location: a particular temperature and a certain humidity. Volumes of air with such distinct features are known as air masses, and the areas in which they form are known as source regions. If an air mass begins to move away from its source region, it initially retains its particular characteristics, but these may become greatly modified with time and distance by the characteristics of the surface over which it moves. As might be expected, air masses have an extremely great influence on the weather of neighbouring areas.

There are four distinct sources, classified on the basis of temperature:
- arctic and antarctic (A)
- polar (P)
- tropical (T)
- equatorial (E)

Antarctic air is sometimes designated 'AA' if it needs to be differentiated from arctic air.

There are two broad categories, continental (c) and maritime (m), depending on whether the air masses develop over land or the oceans and, as might be expected, they have distinctly different humidities. The principal categories of air mass are:

- continental arctic (cA)   extremely cold and dry
- continental polar (cP)   cold and dry
- continental tropical (cT)   hot and dry
- maritime arctic (mA)   extremely cold and humid
- maritime polar (mP)   cold and humid
- maritime tropical (mT)   warm and humid
- maritime equatorial (mE)   hot and humid

Equatorial air is always hot and humid and never continental in nature. Continental arctic air is present over the Arctic only in winter, but persists throughout the year over Antarctica.

Fronts

The boundary between two air masses with different temperatures and humidities is known as a front. There are three types: warm, cold and occluded (pp.70–2). The last will be described in detail when discussing depressions (pp.66–74), but the first two concern us here.

It is important to note that it is the relative temperatures of the two air masses that are important. A front occurs, for example, where arctic (or antarctic) air is in contact with polar air. During the winter in the northern hemisphere, there is often an Arctic Front, stretching from Greenland to Scandinavia, between frigid maritime arctic air and cold maritime polar air. A similar front (also known as the Arctic Front) occurs between continental arctic air and continental polar air over Canada. There is a semi-permanent Antarctic Front, lying between 60–65°S, at the boundary between the cold air flowing from the continental interior and the maritime polar air farther north.

The Polar Front

The two Polar Fronts are important atmospheric boundaries, where many of the highly changeable weather systems that affect temperate latitudes are generated. The boundary between the cold polar air and the warm air from the subtropics is by no means straight, but instead snakes its way around the globe, usually with four or five lobes where, at any given latitude, the cold air has advanced towards the equator, and the warm air towards the pole (overleaf). The lobes consist of a succession of warm and cold fronts, where the warm and cold air, respectively, are advancing.

There are continual changes along the boundary as the lobes grow and shrink. They tend to move eastwards round the Earth, but sometimes become 'blocked' – remaining stationary for considerable lengths of time – or even, occasionally, move westwards. The boundary is rarely stable, and

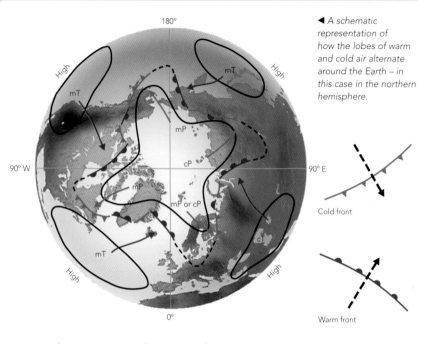

◀ *A schematic representation of how the lobes of warm and cold air alternate around the Earth – in this case in the northern hemisphere.*

Cold front

Warm front

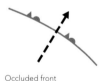

Stationary front

Occluded front

secondary waves repeatedly grow into depressions (low-pressure systems) and the associated anticyclones (high-pressure systems) that bring such changeable weather to the temperate zones. These systems are important for understanding of the weather and will be discussed in detail later (pp.66–74).

Jet streams

Jet streams are fast-flowing ribbons of air in the upper troposphere and lower stratosphere that develop where there are particularly strong contrasts in temperature. Although there are various jet streams, the two most important for surface weather occur where there are major breaks in the tropopause (pp.11–12): near the Polar Fronts, where cold polar air encounters warm subtropical air; and near the subtropical highs, where relatively hot air from the tropics meets cooler air returning from temperate latitudes.

Jet streams form, like winds at lower levels, where there are major differences in temperature at an altitude of (say) approximately 9,000 m (30,000 ft). (See box opposite for a note on the description of heights in the atmosphere.) The column of warm air extends to a much greater height than the neighbouring cold air. Because pressure is determined by the amount of air above a particular location, the pressure on the warm side of the front is greater than that on the cold side, creating a pressure difference. In the northern hemi-

sphere, for example, the warm air (and higher pressure) lies to the south, so the pressure gradient pushes air north. But the Coriolis effect turns the air towards the right (i.e., to the east), giving rise to the strong westerly jet stream. In the southern hemisphere, of course, the pressure gradient forces air south, but as the Coriolis force acts in the opposite direction, the result is still a strong westerly jet stream.

Just like the Polar Front, the jet streams snake their way around the globe, but they also vary in altitude and strength. They are typically thousands of kilometres long, hundreds of kilometres wide, but just a few kilometres deep. Technically, a flow is defined as a jet stream when the minimum velocity is approximately 90–110 kph (56–68 mph). The highest recorded speed is 656 kph (408 mph), found over the Scottish Outer Hebrides on 11 December 1967, but wind speeds vary greatly, often sinking below the lower threshold, so there are breaks in the flow around the Earth. Occasionally a stream will split, giving two branches, and such conditions can have a dramatic effect on the weather below them.

▲ *The Polar Front is clearly visible, running north to south on the satellite image of Western Europe.*

Wind speeds in jet streams are so high that they may make a great difference to the speed of aircraft flying within them. (This was a major factor in their discovery in the 1940s.) A flight travelling in the same direction as the jet stream (generally towards the east) will experience a much greater speed over the ground, whereas flights in the opposite direction will take much longer.

Because jet streams are relatively narrow, but moving at high speeds, they are sites of considerable wind shear, and the turbulence that accompanies it. Aircraft entering or leaving jet streams may encounter severe turbulence, known as clear-air turbulence (CAT). Although

Heights in the atmosphere

Because the use of feet has been accepted as an international standard by the aviation industry, the heights of clouds are usually given in feet, rather than in the SI (metric) units normally used in meteorology. Two approximate equivalents that are important for the description of cloud heights are:

6,500 ft	2,000 m (2 km)
20,000 ft	6,000 m (6 km)

modern forecasting techniques enable most such locations to be predicted, they may still occasionally be encountered unexpectedly. Luckily, although they may be uncomfortable and on rare occasions lead to injury of passengers or crew, they are not a major safety hazard, unlike turbulence near the ground.

Vertical wind shear occurs when layers at two different heights are involved, such as above and below a jet stream, or when a strong warm wind is blowing above a calm layer of cold air at the surface. Horizontal wind shear occurs when there are major changes at one particular level. This occurs

Wind shear

Wind shear develops where there are differences in wind strength or direction (or both) over short distances. If air at two slightly different levels is moving at different speeds (or different directions), wind shear and the associated turbulence occur between them. This is often the cause of billows (p.37) in a layer of cloud, for example. Perhaps surprisingly, occasionally winds in adjacent layers may be blowing in almost opposite directions.

► *Jet-stream cirrus has a highly characteristic appearance that allows it to be recognized easily from the ground. The cirrus streaks may lie along the wind, or appear as billows at right-angles to the flow.*

▼ *Jet-stream cirrus clouds over the Red Sea, photographed from the Space Shuttle.*

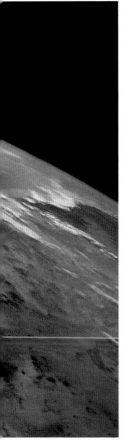

at the sides of a jet stream, or at the edges of strong down-draughts and updraughts within active storm systems.

In general, jet streams exert considerable influence on the development, movement and decay of depressions. As the jets snake around the world at an upper level, the flow of air tends to spread out (diverge) at certain points and become more concentrated (converge) at others. Where the flow diverges it tends to drag air up from lower levels and thus intensifies any depression beneath it. Conversely, where the air converges, some is forced downwards and may prevent a depression from forming or accelerate its decay.

The positions of the jet streams are partially affected by the height of the land surface beneath them. In the northern hemisphere, the North American Rockies, the high Tibetan plateau and the Himalaya are particularly important. The onset of the summer south-west monsoon usually begins when the jet stream switches from lying south of the Himalaya and the Tibetan plateau to flowing on the northern side. Jet streams frequently contain cirrus clouds (p.44), which may be observed from the ground (above) and are often visible in images returned from space (left).

CLOUDS

Despite all the advances in weather forecasting, much can be told about forthcoming weather by observations of the clouds and, in particular, by being able to recognize the different types. Some advice on observing and photographing the sky and on estimating angles – which is useful both for recognizing cloud types and certain optical phenomena – is given later on pp.168–169.

For some reason, people tend to think that recognizing clouds is difficult, but it need not be. Although it is true that no two skies are the same, there are only ten basic types of cloud and they all have distinct characteristics. If you can recognize ten different flowers, birds or cars, why should ten different types of cloud be any more difficult? First, however, it helps to have some idea of how clouds are formed and that involves understanding the remarkable properties of water.

The properties of water
Water is all-important for determining the weather because it exists in three forms (water vapour, liquid water and ice) at temperatures that readily occur on Earth. Liquid water and ice are very familiar to everyone, but water vapour is not

immediately obvious, because it is an invisible gas. If you look at a vigorously boiling kettle, there is a clear gap between the spout and the steam, and this is where water is in the form of gaseous water vapour. The white steam arises farther away from the spout where the water vapour condenses into minute droplets of liquid water.

An important property of water vapour and liquid water is that they contain latent heat. This is best explained by imagining that you have a quantity of ice. It takes heat to melt the ice and turn it into water and additional heat to evaporate the water and turn it into water vapour. When water vapour condenses into water droplets or liquid water freezes, that latent heat is released. This has important consequences for the formation of clouds and other processes that occur in the atmosphere. The release of latent heat is, for example, the reason why plants may be protected from mild frosts by spraying them with water before the temperature drops. When the water droplets on the plants freeze, they release their latent heat and prevent the surfaces of the plants themselves from freezing.

▼ A layer of altocumulus clouds and a sun pillar at sunset.

Cloud droplets and ice crystals

The formation of water droplets and ice crystals are important processes in the formation of clouds (and mist and fog). Condensation of water vapour into droplets of liquid water is relatively straightforward, even though it requires the presence of condensation nuclei – tiny particles of dust or salt on to which the water vapour condenses. Suitable condensation nuclei are present in vast numbers throughout the atmosphere, so condensation occurs whenever humid air is cooled below a temperature known as its dewpoint. The actual temperature at which this occurs depends on the humidity of the air.

Freezing also requires suitable nuclei, but in this case they are not as readily available, because they need to be a specific size and shape. Natural freezing nuclei are often certain minerals derived from clays.

In the absence of suitable nuclei, cloud droplets may exist as liquid water at temperatures far below the normal freezing point of 0°C (32°F), remaining liquid as low as –40°C (–40°F). Such water is said to be 'supercooled', and often exists in the atmosphere. It will freeze spontaneously, even in the absence of suitable freezing nuclei, at temperatures below –40°C (–40°F), and also on contact with solid objects, such as the wings of an aircraft or the leaves of a tree.

The formation of clouds

The majority of clouds (and mist and fog) all form in basically the same manner when humid air is cooled below its dewpoint. This occurs when air comes in contact with a cold surface or when it is forced to rise, when the reduction in pressure causes it to expand and cool. The first process often occurs at night, when the surface cools by radiating heat away to space and the air immediately above it is cooled below the dewpoint, giving rise to mist or fog. It may also occur when air moves over a cold surface, such as one covered in ice or snow.

In the second process, air may be forced to rise in the atmosphere by four different mechanisms:

- through heating of the ground during the day, which creates bubbles of warm air that break away from the surface as 'thermals', in the process known as convection;
- through being forced to rise over a mountain range or similar barrier (p.93), where the mechanism is known as orographic uplift;
- through uplift at frontal surfaces in a depression: frontal uplift;
- through a process known as convergence, where air flows into an area from different directions. When this occurs at the surface, there is only one way in which the accumulation of air can escape, and so it is forced to rise. This process typically occurs at the centre of depressions. Conversely, when convergence occurs at height, the air is forced downwards, giving rise to high pressure at the surface, as found in the subtropical highs and anticyclones in general.

The lapse rate, instability, and stability

Because pressure declines with height, a parcel of air that ascends will expand and cool. (Conversely, a parcel of air that is forced to descend will be compressed and become warmer.) The rate at which this cooling occurs is known as the lapse rate. When air is 'dry', i.e., before condensation sets in – this rate is almost exactly 1 deg C per 100 m. This is known as the Dry Adiabatic Lapse Rate (DALR) and is actually 9.767 deg C per km (17.6 deg F per mi). ('Adiabatic' means that no heat enters or leaves the system: the parcel of air does not exchange heat with its surroundings.) This decline in temperature is, of course, why it is normally much colder at the top of a mountain than at the base. The actual lapse rate within the atmosphere may be less or greater than the amount quoted.

If a parcel of air is warmed, it will begin to rise and will continue to rise all the time that it is warmer than the surrounding air. In other words it continues to rise all the time that its lapse rate is less that than of its environment. Such conditions are said to be unstable, because any vertical movement will continue.

If, however, the environmental lapse rate (ELR, the lapse rate in the surrounding atmosphere) is less than the dry adiabatic lapse rate of the parcel of rising air, a point will come where the parcel of air (which is cooling faster) reaches

Temperatures
To prevent confusion, *actual temperatures* are shown with the degree sign (°) and *differences in temperature* are shown with the abbreviation 'deg'.

the same temperature as its surroundings. Any further rise would cause it to become cooler than the environment and it would sink back. Such conditions are said to be stable. Instability and stability have a major effect upon the types of clouds that form and, as we shall see, these are very distinct.

Naturally, conditions (and the stability or instability of the air) may change with time. Early in the morning, for example, when there has been little heating by sunshine, the wind may force a parcel of air to rise over a range of hills, where it becomes colder than the surrounding air. It is stable, and will sink back towards the ground beyond the hills. Later in the day, however, air near the surface may be strongly heated by the Sun. Now, when the wind carries it up to the top of the hills, it may well remain warmer than the surroundings, and thus continue to rise, rather than sinking back towards the surface. It has become unstable.

Although we have been considering air originating at the ground, similar effects may occur if colder or warmer air moves in at higher altitudes. Relatively warm air above relatively cold air creates stability, and relatively cold air above relatively warm air gives rise to instability.

The effect of water vapour

If the rising air contains water vapour, it will initially behave in the same way as dry air, cooling at the dry adiabatic rate. Eventually, however, it will reach a level at which the humid air reaches the dewpoint. Condensation into cloud droplets occurs, clouds appear and, as described earlier, latent heat is released. The air is now said to be saturated and its lapse rate changes to a lower value because of the additional heat. Depending on the actual temperature of the air, the saturated adiabatic lapse rate (SALR) ranges from about 4 deg C per km (7 deg F per mi) at high temperatures to

▶ *The parcel curve (red) for cumulus cloud developing within a cooler environment (blue).*

DALR: Dry Adiabatic Lapse Rate
ELR: Environmental Lapse Rate
SALR: Saturated Adiabatic Lapse Rate

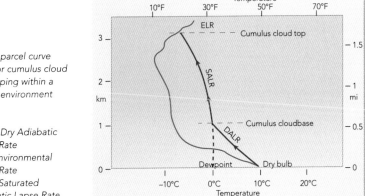

25

about 7 deg C per km (12.6 deg F per mi) at the lowest temperatures down to –40°C (–40°F).

The additional heat available when rising air reaches the dewpoint and latent heat is released may make a dramatic difference to the development of clouds. Instability may increase and the clouds may show rapid growth to much higher levels. Certain types of cloud that exist in the middle layers of the atmosphere are signs of instability at those levels and that major changes may be about to occur. Under certain circumstances, the difference between the dry and saturated lapse rates also has important consequences when air descends.

The environmental lapse rate and inversions

It is important to realize that the environmental lapse rate – as shown by instruments carried aloft by a balloon, for example – may vary considerably and only rarely shows a steady decline in temperature with increasing altitude. At certain levels the temperature may remain steady or even increase with height. Such a level is known as an inversion. Depending on the strength of the inversion, it may prevent a parcel of air from rising any farther, at which point the air and any cloud that has formed tend to spread out beneath it. Similarly, inversions often trap smoke and fumes from industrial chimney stacks in the lowermost layer of air, creating a brownish haze layer that may be visible far downwind (opposite).

Many inversions are relatively shallow with a limited difference in the temperature across them. When convection from the ground surface is vigorous, rising thermals and cloud cells are often strong enough to penetrate through the inversions and continue growing above them. However, the tropopause itself (p.4) is a deep, major inversion, and as such, limits the growth of even the most vigorous cumulonimbus clouds. When the updraughts are exceptionally

▼ *This image, taken from the Space Shuttle, clearly shows how vigorous convection causes the updraughts to overshoot the capping inversion. Image obtained over Brazil.*

▲ Smoke from the chimney of a power station rising as far as an inversion – and causing a small cumuliform cloud – and then spreading as a plume downwind, trapped by the inversion.

strong, they may cause what is known as an overshooting top – a mound of cloud immediately above the vigorous updraught, where the rising air penetrates a short distance into the stratosphere, before falling back and spreading out at a slightly lower level.

CLOUD FORMS

Like plants and animals, clouds are classified by type (genus), species and variety. There are ten types (genera), 14 species and nine varieties, as well as six supplementary features. The ten main types will be described here together with some of the species, varieties and supplementary features that are particularly relevant to understanding current or forthcoming weather. Luckily, the ten main types are fairly easy to recognize (overleaf).

The various genera may be grouped in two different ways. A broad division is into two classes: cumuliform clouds and stratiform clouds. The first includes heaped clouds, such as cumulus and cumulonimbus, which are associated with instability. The second comprises layer clouds, such as stratus and altostratus, which generally arise under stable conditions. A third group is sometimes added: cirriform cloud. These consist of ice crystals and occur at high altitudes.

Meteorologists normally divide the ten types into three groups (high, medium and low), according to the height of their bases:

Cloud classification

Genus	Main, distinctive forms
Species	Cloud structure and shape
Variety	Transparency and arrangement of cloud elements
Supplementary features	Specific forms associated with particular genera or species

27

Cloud types

High
3–18 km
10,000–
60,000 ft
Cirrus

Medium
2–8 km
6,500–
26,000 ft
Nimbostratus

High
3–18 km
10,000–
60,000 ft
Cirrostratus

Low
0–2 km
0–6,500 ft
Cumulus

High
3–18 km
10,000–
60,000 ft
Cirrocumulus

Low
0–2 km
0–6,500 ft
Stratus

Medium
2–8 km
6,500–
26,000 ft
Altocumulus

Low
0–2 km
0–6,500 ft
Stratocumulus

Medium
2–8 km
6,500–
26,000 ft
Altostratus

0.5–18 km
1,600–60,000 ft
Cumulonimbus

High Bases above 3–6 km (c.10,000–20,000 ft)
Medium Bases at 2–6 km (c.6,500–20,000 ft)
Low Bases below 2 km (c.6,500 ft)

For the high and medium clouds, the figures given are average values, because the range of height varies depending on latitude, being generally higher in the tropics and lower towards the poles. One cloud type, cumulonimbus, may extend through all three zones, from just above the surface to the very top of the troposphere.

Cloud heights

	Range of height	Type (Genus)	Abbreviation
High	3–18 km (10,000–60,000 ft)	Cirrus	Ci
		Cirrostratus	Cs
		Cirrocumulus	Cc
Middle	2–8 km (6,500–26,000 ft)	Altocumulus	Ac
		Altostratus	As
		Nimbostratus	Ns
Low	0–2 km (0–6,500 ft)	Cumulus	Cu
		Stratus	St
		Stratocumulus	Sc
	0.5–18 km (1,600–60,000 ft)	Cumulonimbus	Cb

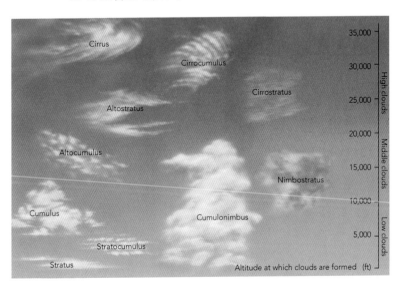

Cumulus (Cu)

Cumulus clouds are very familiar and are the type that most people imagine when the word 'clouds' is mentioned: fluffy heaps of cloud with rounded tops and flat bases. Cumulus clouds arise from thermals, heated bubbles of air that have broken away from the surface and then rise, invisible, until reaching the condensation level. Early in the day, when heating of the ground is just beginning, the thermals are small, and any clouds are usually ragged fragments, without any clearly defined bases. This form is known as cumulus fractus (below).

Such wisps of cloud normally have a limited lifetime and soon fade away. This decay arises because a thermal is not completely isolated from its environment and its circulation tends to draw drier, cooler air in from the surroundings. Generally, the smaller the thermal (and resulting cloud), the quicker the mixing and decay occur.

Gradually, as heating by the Sun increases and the thermals become stronger, larger clouds appear, with the characteristic rounded tops and flat bases, which give a clear indication of the height of the condensation level. On occasions, under relatively stable conditions, the clouds have a distinctly flattened appearance – they are known as cumulus humilis (opposite, above) – and this species commonly occurs ahead of a warm front (p.69), when warm air is extending overhead, preventing further upward growth.

Small cumulus may continue to grow until their vertical height is approximately the same as their horizontal extent.

▶ *A collection of cumulus humilis clouds as convection increases on a fine day. There are already signs that some clouds are developing into cumulus mediocris.*

◀ *Wisps of cumulus fractus, forming early in the day, and rapidly decaying, to be replaced by larger clouds as convection increases.*

This species is known as cumulus mediocris (below) and, like cumulus fractus and cumulus humilis, it does not give rise to any rain.

▼ *Cumulus mediocris clouds, with several showing signs of active growth.*

The next stage in growth produces clouds that are distinctly higher than they are wide. This species, known as cumulus congestus (overleaf), may be a source of rain,

CLOUDS

▲ *Cumulus congestus (towering cumulus) over the sea. The main cloud shows a fine display of crepuscular rays radiating from the position of the Sun.*

Cumulus species

Cumulus fractus	Ragged wisps of cloud
Cumulus humilis	Flattened 'pancakes' of cloud
Cumulus mediocris	Clouds of equal width and height
Cumulus congestus	Clouds higher than they are wide

because relatively strong turbulence and great vertical extent cause the tiny cloud droplets to collide and eventually grow into raindrops. Meteorologists describe these clouds as 'warm clouds' that give rise to 'warm rain'. Rain-bearing cumulus congestus clouds are frequently encountered in the tropics (where they are a common source of rain throughout the year), and also occur in summer at temperate latitudes.

Cumulonimbus (Cb)

At temperate latitudes most of the rain from cumuliform clouds is produced by cumulonimbus clouds (opposite), which are, in effect, the final stages in the growth of cumulus. These may be called 'cold clouds' and their rain 'cold rain', because they grow so high that the water droplets freeze, turning into ice crystals. The change from a cumulus cloud to a cumulonimbus is marked by a change in the appearance of the top of the cloud: the hard 'cauliflower-like' outline becomes softer, and then takes on a fibrous appearance. These stages are actually known as cumulonimbus calvus (Latin: 'bald') and cumulonimbus fibratus (Latin: 'fibrous').

As mentioned earlier (p.23), ice crystals form only in the presence of suitable freezing nuclei, and because these are often absent from the air, cumulonimbus clouds may grow extremely high before freezing – known technically as 'glaciation' – sets in. Once freezing occurs, the ice crystals grow rapidly and are soon too heavy to be supported by the updraughts within the cloud. When they reach warmer air at lower levels, they melt and become raindrops. In winter, cumulonimbus clouds may be quite shallow and give rise to snow, rather than rain

Isolated and relatively small cumulonimbus form the 'showers' often described by forecasters. Such small showers have a limited life: their growth stage is about 20–30 min-

▲ *A vigorous cumulonimbus cloud. The sharply defined outline to the rising cells in the foreground shows that the cloud is still actively growing. There are suggestions that rain is falling in the background on the right-hand side.*

▼ *Spectacular anvils on two cumulonimbus clouds, illuminated by the low Sun. The more distant anvil is older and has spread out to a greater extent.*

▲ *Mamma beneath the anvil of a large cumulonimubus cloud, thrown into striking relief when lit by the low Sun.*

utes, followed by a mature stage with rain for a similar period of time. The falling rain tends to destroy the updraughts, which collapse, and the cloud takes another 20–30 minutes to decay. The overall lifetime of such individual 'cells' is therefore between one and one-and-a-half hours. Cumulonimbus may, however, grow into very substantial clouds, consisting of clusters of cells and become correspondingly stronger. They may become major storm centres, producing lightning, hail and even, in extreme cases, giving rise to tornadoes. Such storm systems will be described later (p.79).

Cumulonimbus frequently reach the tropopause, which, being an inversion, arrests their upward growth. The rising air spreads out beneath the tropopause to form the characteristic 'anvil' shape (known as cumulonimbus incus, p.39). Such anvils frequently form giant cirrus plumes that may be carried far ahead of the approaching storm. Frequently heat is lost to space from the top of the anvil, and cold air descends below the overhanging cloud, cooling the air below the dewpoint and giving rise to hanging pouches of cloud, known as mamma (Latin: 'breast' or 'udder'), or long, contorted tubes of cloud. These may appear extremely striking when illuminated by the low Sun.

Stratus (St)

Stratus is another common cloud, grey or blue-grey in colour, which often extends as an almost featureless layer right across the sky. It frequently hides the tops of buildings and blankets high ground, and to anyone within it appears as mist or fog. It is the lowest of all clouds, and only rarely is its base above 500 m (about 1,650 ft). When the layer is extensive, the base itself is diffuse, rather than ragged, and may show some slight undulations. The tops of hills often protrude above a layer of stratus, and its upper surface tends to be more irregular than the base.

Stratus often develops from overnight fog or mist. These form over low-lying ground when the latter loses heat to space overnight, cooling the air immediately above it below

▲ *Stratus clouds over the Alps, breaking up into stratus fractus with the slowly increasing warmth of the day.*

▼ *Two layers of stratus cloud are seen here, the higher of them hiding the mountain-tops and the lower a thin layer that probably formed immediately above the ground during the night and then rose with the return of warmth in daylight.*

the dewpoint. When heating begins after sunrise, the layer of fog may lift to produce a low layer of stratus cloud, which then tends to disperse as the heating becomes stronger or a wind develops. Stratus is very common in depressions, but being a low cloud it is strongly affected by turbulence when the wind rises. It then breaks up into ragged patches of cloud (known as stratus fractus) and gradually disperses. Such stratus fractus may sometimes be confused with cumulus fractus (p.30), which it superficially resembles, but the conditions under which the two cloud species occur are normally clearly distinct, so they are easily distinguished. Under quiet conditions, a layer of stratus may sometimes lose heat by radiation from its upper surface to space, causing gaps to appear in the layer and for it to become transformed into stratocumulus.

▲ A layer of stratus that is beginning to break up into stratus fractus and reveal the Sun.

Stratus is frequently very thin and allows the position of the Sun – and sometimes even the Moon – to be seen clearly through it. This is known as stratus nebulosus. Because stratus generally consists of water droplets, a corona (p.58) is sometimes visible around the Sun or Moon. On very rare occasions, ice crystals may be present and may produce a halo (p.54). Stratus does not often give rise to any form of precipitation, but just occasionally may produce a fine drizzle or, under cold conditions, a few snowflakes or tiny ice crystals.

Stratocumulus (Sc)

Stratocumulus is an extremely common, low, layer cloud. Satellite observations show that it is present over large areas of the world's oceans where there are relatively quiet, stable conditions with warm air overlying cooler waters. It generally forms through the spreading out of cumulus below an inversion. As the cloud spreads out it thins, so frequently the base of the stratocumulus is distinct and noticeably higher than the condensation level at which the cumulus clouds became visible.

The appearance of stratocumulus is very varied, because it consists of rounded masses or rolls, or flattened 'pancakes' of cloud, with distinct shading. The colours may range from white to grey and dark blue-grey. To be classed as stratocumulus, the individual cloud elements must be more than 5° across, measured 30° above the horizon. This size distinguishes them from the small cloud elements in altocumulus and the much smaller elements in cirrocumulus. (Measuring angles is described on p.169.)

Sometimes the gaps between the individual cloudlets become very narrow or even disappear, leaving an almost unbroken, undulating blanket of cloud. Stratocumulus often appears in long rolls (known as billows), which tend to lie across the direction of the wind at height and which also

▼ A sheet of stratocumulus, showing how the more distant cloud elements tend to give the impression of an unbroken layer.

indicate that the strength or direction of the wind varies slightly with height. In general, however, stratocumulus forms only when winds are light to moderate.

Stratocumulus is frequently found in the warm sector of a depression (p.71), particularly where the air is subsiding gently. It tends to form under fairly quiet conditions when the temperatures do not vary greatly between day and night. The weak thermals are generally unable to break through the overlying inversion. There may be scattered stronger thermals that are able to do so, but these are difficult to recognize from the ground – although occasionally recognizable by their darker, shadowed bases – but can be seen clearly from aircraft flying above the cloud.

The cloud itself tends to reduce maximum daytime temperatures and (by acting as a blanket) increase those at night. It is often very difficult to predict when a layer of stratocumulus will break up. If the day is bright (i.e., the cloud is thin) and there is a strong rise in atmospheric pressure, subsiding air above the cloud causes it to thin and eventually break up. When the stratocumulus is thick and dark, however, subsiding air alone may not be sufficient to dissipate the cloud and the layer is likely to break up only if the wind rises, creating strong turbulence and mixing of the air. Even then, if there is strong heating of the ground early in the day, new thermals may arise that simply fill in the holes that have been created. If the breaks occur later in the day, when heating is declining, strong thermals will not be produced and the cloud may continue to disperse to give rise to a clear night.

Altostratus (As)

The medium-level cloud that corresponds to stratus is altostratus. It, and the higher cirrostratus, often occur in depressions where warm air is rising at the warm and cold fronts. Altostratus is a relatively featureless, grey and

▲ *(Left) A thick layer of stratocumulus, photographed from the air. The gaps between the cloud elements are small, and little blue sky would be visible from the ground.*
(Centre) A fine display of stratocumulus billows (stratocumulus undulatus). The wind at altitude is flowing approximately at right-angles to the billows.

▶ *A typical sheet of thin altostratus, showing few features except for some striations running downwind. The position of the Sun would not be visible in thick altostratus.*

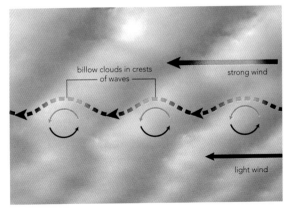

billow clouds in crests
of waves

strong wind

light wind

▲ Formation of billows.

uniform layer cloud, although it may occasionally show
a fibrous or striated structure. It may consist of water
droplets alone but is more frequently a mixed cloud with
both water droplets and ice crystals. When altostratus is
thin, the position of the Sun is clearly visible, looking as if
seen through ground glass, and it may be surrounded by a
corona (p.58). The edges of an altostratus sheet often show
iridescence (pp.58–9). Thick altostratus will completely
obscure the Sun and may appear extremely dark. It causes
diffuse light at ground level, where objects no longer cast
distinct shadows.

Altostratus frequently produces large amounts of precipitation in the form of rain or snow, but this does not always

reach the ground, evaporating instead in the drier air beneath the cloud. The precipitation often appears in the form of virga (p.43). It may cool a lower layer of air and raise its humidity to such an extent that turbulence may cause ragged fragments of an accessory cloud known as pannus (actually stratus fractus) to appear. Very occasionally, the fragments of stratus may merge to give a more or less continuous layer of stratus.

Sometimes precipitation from an invisible layer of altostratus may seed a lower layer of stratus and cause drizzle or very light rain at the surface. At the warm front of a depression altostratus commonly thickens and lowers and becomes nimbostratus, which may generate prolonged periods of heavy rain. Conversely, at the cold front, the nimbostratus may thin and lift and turn into altostratus.

In the tropics, active cumulonimbus clouds may spread out at an inversion and give rise to extensive layers of altocumulus that trails behind the active system (or may even, on occasions, precede it). At higher latitudes, similar active storm systems generally produce stratus and stratocumulus rather than a layer of altostratus.

Nimbostratus (Ns)

Because of its close association with altostratus, it is sensible to discuss nimbostratus here. It is a grey, often very dark grey, layer cloud that produces large quantities of rain or snow. Nimbostratus always displays these features, but otherwise shows little variation, in contrast to most other cloud types, and it has no species or varieties.

Nimbostratus is a dense cloud that always hides the Sun and Moon, so it does not give rise to any optical phenomena. The precipitation falling from its base means that this normally appears diffuse, rather than sharply defined. As with altostratus, precipitation frequently cools the air beneath nimbostratus and increases its humidity, leading to the formation of ragged patches of pannus (stratus fractus), which often hide the base of the nimbostratus itself.

Nimbostratus is most commonly found in association with a depression, when it is normally the last member of the succession: cirrostratus, altostratus and nimbostratus. Although it is generally regarded as being associated with warm fronts, it is also found at cold and occluded fronts (p.71). At the last, it may sometimes produce days of persistent heavy rain or snow, with resultant flooding or major snowfalls. When associated with a depression, the precipitation from nimbostratus tends to give bands of heavy rain, separated by breaks of less (or no) precipitation. These bands of precipitation run roughly parallel to the fronts and are particularly marked at the warm front.

On rare occasions, nimbostratus may arise when altocumulus or stratocumulus thickens into a heavy layer of cloud.

Similarly, it may sometimes be created by active cumulonimbus systems or even cumulus congestus, but then only a relatively small area is affected, unlike the vast areas it covers in depressions.

Altocumulus (Ac)

Altocumulus may produce some striking and beautiful skies. This is mainly because it shows a wide range of different forms. There are four different species and seven varieties – more than any other cloud type. The clouds are generally similar to the lower stratocumulus and the higher cirrocumulus in that they consist of layers of individual small cloudlets. These are generally paler than stratocumulus elements, being white or grey with some shading, which is not as strong as that found with stratocumulus. (Cirrocumulus shows no shading.) The size of the cloud elements lies between 1° and 5°, measured 30° above the horizon. (A simple method of measuring angles is given on p.169.) Because of the clouds' higher altitude than stratocumulus, the gaps between the individual cloudlets are easier to see, and blue sky is usually visible through them. The individual elements are often arranged in striking billows.

Altocumulus varies considerably in its thickness. It is sometimes thin enough for light from the Sun or Moon to shine through it. When it consists solely of water droplets – although these are often supercooled – it may give rise to a corona (p.58) or iridescence (pp.58–9). At other times it may contain ice crystals and give rise to various halo effects, such as a sun pillar (p.57) or a parhelion (p.55).

▲ *Dark nimbostratus clouds over Kinder Scout in Derbyshire, seen during a short break in a long period of almost continuous rain. Below the main layer, some ragged cloud (known as pannus) has formed where rain has cooled the air below the dewpoint.*

▲ A fine sheet of moderately high altocumulus. The individual elements are more than 1° across and do show some signs of shading, otherwise they would be considered cirrocumulus.

◀ Clearly defined altocumulus billows.

▶ A very typical sheet of altocumulus stratiformis.

▲ *Altocumulus clouds with well-developed virga. The bend occurs where ice crystals melt into water droplets.*

As with stratocumulus, altocumulus may form through the spreading out of cumulus at an inversion, through the break-up of a layer of altocumulus or nimbostratus or through generalized uplift. It often occurs both ahead of, and behind major cumulonimbus systems. Altocumulus does not generally give much precipitation, but is often accompanied by

virga – trails of precipitation that do not reach the surface, but evaporate in the drier, lower air.

▲ *Small tufts of altocumulus floccus are a sign of instability at cloud height.*

Altocumulus frequently occurs as extensive sheets covering a large part of the sky, and this species is known as altocumulus stratiformis. It often gives rise to spectacular sunsets and sunrises when the layer is illuminated by red or orange light from the Sun. Two species are very indicative of strong convection, heavy showers and possibly severe thunderstorms developing later. These species are altocumulus castellanus (right), and altocumulus floccus (above). Both indicate instability in the middle layers of the troposphere. If convective cloud rises to that height, it will gain additional buoyancy through the release of latent heat (p.23) and rise to much greater altitudes, producing deep, vigorous cumulonimbus clouds. The fourth altocumulus species, altocumulus lenticularis, is one of the family of wave clouds, and will be described later.

Cirrus (Ci)

Cirrus clouds consist of ice crystals and these are often drawn out into long trails across the sky, commonly

▲ *Typical cirrus, clearly showing the generating heads in which the ice crystals are forming.*

known as 'mares' tails'. Cirrus is seen in a variety of situations: ahead of an approaching warm front; as large, and often very dense plumes forming the top of cumulonimbus anvils; as long lines of cloud spreading from aircraft condensation trails; and as loops and whorls of high cloud under fine weather conditions. Quite frequently, highly distinctive bands of jet-stream cirrus may be observed stretching right across the sky.

Although cirrus occurs in all these situations, only in the case of cumulonimbus clouds or thunderstorms is it associated with a sudden change in the weather. In most cases, depending on the exact conditions, it may be an early indicator of a deterioration of the weather. This is particularly the case when cirrus gradually invades the sky from one direction, slowly becoming thicker until it eventually turns into a layer of cirrostratus. This is a sure indication that a warm front is approaching.

There are various species of cirrus of which two, cirrus uncinus (hooked cirrus, above) and cirrus fibratus (fibrous cirrus, overleaf) are the forms that often thicken into cirrostratus. Cirrus spissatus (dense cirrus) is very thick and appears dark grey against the light. It forms the dense heads of cumulonimbus clouds and frequently remains behind when the rest of the cumulonimbus has decayed.

◄ *A field of altocumulus castellanus, where the obvious growth of the upper surface is a clear indication of instability at that height.*

Frequently cirrus consists of a small tuft of cloud (the generating head) where freezing is taking place, followed by a trail of ice crystals. This is most noticeable in the species known as cirrus uncinus. The trails may be considered as a form of virga (pp.43–4), and indeed it could be said that

▲ Tangled cirrus (cirrus intortus) as seen here often accompanies relatively quiet weather, including anticyclonic situations.

▼ Cirrus fibratus, together with the edge of a patch of thicker cirrostratus at lower right.

cirrus clouds are nothing but virga. Cirrus is a high cloud, so the trails of ice crystals rarely reach the surface, and then only at high latitudes and under cold conditions.

Cirrostratus (Cs)

Cirrostratus is usually a thin sheet of cloud, which often goes unnoticed as it steals across the sky. Frequently people do not realize it is there until it begins to reduce the heat of the Sun. When they raise their hands to block out the sunlight, they are frequently surprised to see a halo around the Sun. Halo phenomena (p.54) are very common in thin cirrostratus and are visible approximately once every three days at middle latitudes. Unlike altostratus, cirrostratus is always thin enough for the Sun to cast shadows.

▲ *Cirrus uncinus gradually filling the sky ahead of a warm front.*

Cirrostratus may be a featureless sheet of cloud (cirrostratus nebulosus), but often shows a fibrous structure (cirrostratus fibratus, overleaf), particularly when it has arisen through the gradual increase and thickening of cirrus cloud. It is generally visible ahead of an approaching depression when, ahead of the warm front, it forms part of the sequence: cirrus, cirrostratus, altostratus and nimbostratus.

◀ *A layer of extremely even cirrostratus together with a 22° halo, which has a distinct, inner red edge.*

47

◄ *The edge of a cloud layer where individual streaks of cirrus are merging to give a sheet of cirrostratus fibratus.*

Although most commonly associated with depressions, cirrostratus may sometimes be created by cumulonimbus clouds or be the remnants of cirrus plumes. Very rarely it may result from the decay of altostratus. Of the two varieties, the most common is cirrostratus undulatus, when it takes the form of a series of waves. These, however, are difficult to see unless the layer of cloud is illuminated by either the rising or setting Sun.

Cirrocumulus (Cc)

The third cirriform cloud, cirrocumulus, generally resembles stratocumulus and altocumulus in that it consists of a sheet of individual cloud elements. These appear, however, much smaller in cirrocumulus, being less than 1° across, measured 30° above the horizon (a guide to measuring angles can be found p.169). The cloud elements are white or pale blue, thin and do not show any signs of shading, so the layer of cloud may be difficult to see, showing little contrast against the background sky.

Cirrocumulus consists of ice crystals, sometimes mixed with supercooled water droplets, although the latter normally rapidly freeze. The positions of the Sun and Moon can

▲ *An unusually well-defined layer of cirrocumulus, merging into thick cirrostratus in the distance.*

always be seen clearly through this thin cloud, which may display optical phenomena such as iridescence (pp.58–9) and coronae (p.58).

Cirrocumulus may sometimes occur when shallow convection sets in within a sheet of cirrostratus. It may also remain behind when altocumulus decays. Although the most common species is cirrocumulus stratiformis (sheet cirrocumulus), it displays many forms in common with altocumulus, including cirrocumulus castellanus (turreted cirrocumulus), and cirrocumulus floccus (tufted cirrocumulus). It may also occur as a wave cloud (p.90), and one variety, cirrocumulus undulatus, is moderately common, with the cloud elements arranged in regular rows.

Wave clouds

When the wind encounters a range of hills or mountains, the air will be set into a series of waves above and stretching downwind of the obstacle. When conditions are right, cloud will form in the peaks of the waves, where the temperature drops below the dewpoint, and dissipate in the troughs where the air descends and warms. Unlike all other clouds, which move with the wind, wave clouds – also sometimes called lee-wave clouds – remain stationary for hours on end and alter or disappear only when the wind changes strength or direction (p.90). The clouds themselves are constantly changing, however, forming on the upwind edge and dispersing on the downwind side. (You can sometimes see this happening if you examine such clouds with binoculars.) Gen-

▲ *A giant altocumulus lenticularis cloud, caused by uplift over a range of mountains, hidden by lower stratocumulus clouds.*

▼ *Trains of wave clouds over the city of Venice. The crests and troughs of one train are clearly visible.*

▲ *A 'pile d'assiettes', clearly showing the existence of multiple thin layers of humid air.*

erally, the crest of a wave lies parallel to the line of hills or mountains, but sometimes a relatively isolated peak may produce an elongated train of clouds, with wave-like structure that stretches far downwind.

Wave clouds are also known as lenticular clouds because of their lens- or almond-like form. The species most commonly seen is probably altocumulus lenticularis, but both stratocumulus lenticularis and cirrocumulus lenticularis also occur. Although the '-cumulus' ending might be taken to indicate that they arise through convection as do other cumuliform clouds, in fact they form in stable layers of humid air. Frequently, there is a series of humid layers, one above the other, and each layer may produce a lenticular cloud or clouds. These stacks of wave clouds are known by the term '*une pile d'assiettes*' (French: 'a pile of plates').

Unusual clouds – nacreous clouds

Although they have no direct implications for the weather, occasionally beautiful displays of nacreous (or 'mother-of-pearl') clouds occur, and these are usually so striking that when they do they are mentioned on radio and television news programmes. These clouds (known technically as polar

stratospheric clouds) form at altitudes of 15–30 km (9–18 mi). They consist of ice particles which diffract the light in a form of iridescence (pp.58–9). Although a white form does exist – most commonly seen from Antarctica – it is their pastel colours that make them particularly striking. As with iridescence, the bands of colour correspond to areas of the cloud where all of the particles are the same size.

The clouds are visible from high latitudes (beyond 50°N and 50°S, approximately), when the Sun is below the observer's horizon, but the clouds are still illuminated, i.e., they are seen after sunset and before sunrise, although the early morning displays are less frequently recorded. Nacreous clouds are a form of high-altitude wave cloud and tend to be recorded when there are high-speed upper-level winds and in the vicinity of deep depressions.

▲ *Nacreous clouds in the lowest portion of the stratosphere. They are beginning to lose their delicate pastel shades, because they are illuminated by red and orange light from the setting Sun.*

Unusual clouds – noctilucent clouds

Even rarer than nacreous clouds are noctilucent clouds (NLC), which are visible in the middle of the night in summer, but only from latitudes between approximately 45° and 60° either side of the equator. At these times, the Sun is below the poleward horizon, but still illuminates the clouds, which are the highest in the atmosphere, at an altitude of about 80–85 km (50–53 miles). They lie just below the mesopause – the coldest point in the atmosphere (p.6).

▲ A bright display of noctilucent clouds, seen in the north around midnight. The billow-like structure is highly characteristic of these high-altitude clouds.

The clouds have a characteristic silvery-white appearance and are generally visible for an hour or so around midnight. They display structures somewhat similar to those of cirrus clouds and are often mistaken for them. Frequently, lower clouds may be seen silhouetted against them. They consist of ice particles in very thin sheets and it is the folds and waves in the thin sheets that give rise to the apparent structure. Perhaps surprisingly, they seem to be extremely high-altitude wave clouds, tenuously linked to mountain ranges far below them. The clouds are carried towards the west or south-west by upper winds, but the structures tend to move in the opposite direction.

There are two particular mysteries about noctilucent clouds. Because of the structure of the atmosphere, water vapour cannot be transported from the surface into the region of the mesopause, where they occur. It is suspected, therefore, that the water found in the ice crystals of noctilucent clouds is possibly cometary water that arrives from interplanetary space. The second mystery is that there are no positive records of NLC before 1884, when the sky was under detailed scrutiny following the catastrophic eruption of Krakatau in 1883, and NLCs were first noted. There are indications that they have also become more frequent in very recent decades and some displays have been positively linked with rocket and satellite launches.

OPTICAL PHENOMENA

Many different optical phenomena are visible in the atmosphere, but only a few are directly related to current or forthcoming weather, so only a selection of them are described here. They may be roughly divided into phenomena seen towards the Sun (including haloes), and those seen on the opposite side of the sky (such as rainbows).

Haloes
There is a large number of different halo phenomena – far too many to be described here – all of which arise through the reflection or refraction of light by ice crystals. Some are most commonly seen in association with cirrostratus ahead of a warm front, and for this reason suggest that the weather will gradually deteriorate.

The most common effect is the 22° halo, i.e., one with a radius of 22°, which often appears as a complete circle surrounding the Sun (p.47) or Moon, but sometimes exhibits breaks where the structure of the cirrostratus changes. When faint it is colourless, but when strong displays a reddish tint on the inner edge with yellow farther from the centre. On rare occasions the outer circumference may appear bluish.

▼ A 22° halo, together with two bright parhelia, both showing the tail that points away from the Sun, photographed in Antarctica.

The area within the ring usually appears darker than the surrounding sky. A second halo, with a radius of 46° is sometimes visible, although rarely complete, and normally very much fainter than a 22° halo.

▲ *A particularly strongly coloured parhelion in a patch of cirrus cloud.*

Parhelion

A parhelion (pl. parhelia), also known as a mock Sun or sun dog, is a bright spot, often showing spectral colours, created by light being refracted within hexagonal ice crystals. It lies

◄ *A bright parhelion with a tail extending along the parhelic circle: a white arc, sometimes visible running parallel to the horizon at the altitude of the Sun.*

at the same height above the horizon as the Sun, close to the position of the 22° halo, and frequently shows a white tail extending away from the Sun. Parhelia often occur in pairs – one on either side of the Sun – and are frequently visible in wisps of cirrus cloud even when there is no halo. When they occur in a thin sheet of gradually thickening cirrostratus, they, like haloes, indicate that a depression is approaching.

Circumzenithal arc

A circumzenithal arc is a bright band of spectral colours that forms part of a circle centred on the zenith (the point direct-ly above the observer's head). Its actual radius depends on the altitude of the Sun: it has the smallest radius, and its colours are faintest, when the Sun is low on the horizon. Its length also varies, reaching a maximum of 108° when the Sun's altitude is 15°. Its largest radius occurs when the Sun is 22° above the horizon, and it then touches the top of the 46° halo (which may, or may not, be visible). The colours are strongest with solar elevations of 15–22°.

Circumzenithal arcs are not as common as 22° haloes, but are more often seen than 46° haloes. A similar band of even stronger colours, known as the circumhorizontal arc, runs between 58° and 80°, and consequently is visible only at midsummer from southern Europe and the northern USA, and not from more than 55°N or S.

▼ A brilliantly coloured circumzenithal arc in cirrus gradually thickening ahead of a depression.

Sun and Moon pillars

Halo phenomena are governed by the shape of the ice crystals that happen to be present. When there are large numbers of flat, hexagonal, plate-like crystals they often float in the atmosphere with their flat surfaces parallel to the ground. Sunlight, or moonlight, is reflected by the flat surfaces and appears as a column of light above and below the Sun. Such sun (or moon) pillars are moderately common, but are sometimes confused with a bright crepuscular ray. Generally, however, it is possible to determine the type of cloud that is present and only cirriform clouds will give rise to sun pillars. From an aircraft, or a high vantage point, the pillar is sometimes visible below the Sun. Under ideal conditions (which occur only rarely) it is possible to see that the sides of the pillar are not perfectly straight, and that the sun pillar is in the shape of an extremely elongated figure of eight, centred on the Sun.

▶ A sun pillar in cirrostratus undulatus at sunset.

Corona

A corona consists of a coloured disk, generally with a brownish-red outer edge, that closely surrounds the disk of the Sun or Moon. (Coronae are more commonly noticed around the Moon, because they are easily hidden by the glare of light from the Sun.) This inner disk, or aureole, may be just 2–3° in diameter, but is not always visible. Generally, the Moon is surrounded by one or more (often partial) larger rings exhibiting spectral colours. The colours in coronae do not arise through refraction through cloud particles but because of diffraction – the light is bent through various angles around the individual particles. The purer the colours, the smaller the range of sizes of the particles. Coronae occur when the cloud particles (either water droplets or ice crystals) are smaller than normal. When seen on a foggy day, a corona indicates that the fog is thinning and will soon lift. Coronae may also appear when the air is full of pollen. They are, for example, fairly commonly seen when the pine forests of Scandinavia or North America are releasing large quantities of pollen into the air.

▲ A solar corona, where the Sun itself is hidden behind a rock spire on the side of a mountain. In the original at least three sets of coloured rings may be seen.

Iridescence

Bright bands of colour are frequently visible around the edges of clouds that are about 35–40° from the Sun, but these (like coronae) often go unnoticed until the Sun's disk is hidden. Like coronae, iridescence arises from diffraction and, similarly, the purer the colour, the more even the distribution of particle sizes. Iridescence is common in altocumulus,

▲ *Delicately coloured bands of iridescence in mixed cirrocumulus and cirrostratus clouds.*

▼ *Strong iridescence in the exhaust trail of a research rocket, launched from the western United States.*

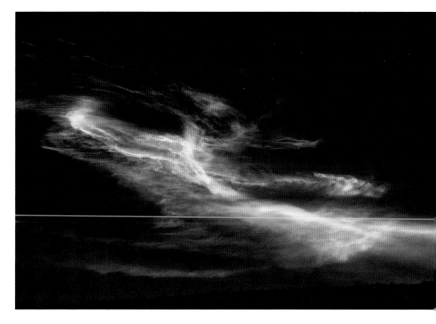

and generally indicates slow uplift of the air and some middle-level instability, although it may also be seen in other types of cloud. It is the cause of the beautiful colours seen in nacreous clouds (p.52) and often appears in the trails left by high-altitude rockets (p.59).

Rainbows

Of the optical phenomena seen on the opposite side of the sky to the Sun, rainbows are, of course, the most common, and have been seen by practically everyone. They are created when light from the Sun, or Moon, falls on raindrops. The most frequent form occurs as a circular arc, centred on the anti-solar point – the point on the celestial sphere directly opposite to the Sun, relative to the observer's head. (So even people standing next to one another actually see slightly different rainbows.) The higher the Sun is in the sky, the lower the top of the bow will appear. It forms a perfect semicircle if the Sun is on the horizon. From an aircraft, however, if conditions are right, rainbows may appear as perfect circles.

▼ *A fine double rainbow, photographed in Mongolia, clearly showing the darker Alexander's Dark Band between the primary (inner) and secondary bows.*

The most common rainbow, with the strongest colours, is known as the primary bow and has a radius of 42°, with red on the outside and violet on the inside. It arises when light from the Sun (or Moon) undergoes a single reflection inside the individual raindrops. Another, fainter, secondary bow, with reversed colours, is quite frequently seen with a radius

▶ A strong primary bow and a reflected rainbow, photographed over a lake in Finland. The surface of another lake, producing the reflected light, lay behind the observer.

of 51°, and this arises through a double reflection within the raindrops. In between the primary and secondary bows there is a distinctly darker area of sky, known as Alexander's Dark Band, where light is reflected away from the observer.

A rainbow is not always a perfect, unbroken arc. Breaks may appear where no rain is falling or where the raindrops are falling within the shadow of other clouds. There is no direct relationship to forthcoming weather, although, given that weather systems tend to move from west to east, if a rainbow is seen in the west early in the day, a shower or rain-clouds are probably approaching, and if one is seen in the east late in the day, the clouds are passing away to the east and the weather may be improving.

The purity of the colours depends strongly on the size of the raindrops. When these are large, red tends to predomi-nate, but with smaller drops the red hue fades and the out-side of the primary bow appears orange. When the droplets are extremely small, the colours tend to merge to give a white bow. Because this is more commonly seen with mist or fog, it is known as a fogbow.

Glories

A glory is a series of coloured rings surrounding the shadow of the observer's head cast on to cloud or mist, i.e., around the anti-solar point. At one time, these were seen only by observers on mountains, looking down on to banks of cloud

or fog, but are now most frequently noted from aircraft. Sometimes the cloud is near enough for the shadow of the plane to be seen, but more frequently the distance is too great for this to be distinguished and the set of rings appears to slide across the surface of the cloud. As with coronae and iridescence, the effect arises through diffraction by cloud particles. Occasionally, a glory will be accompanied by a white fogbow, which, if viewing conditions permit, may be seen as a complete circle, about 42° in radius, surrounding the glory.

▲ A fine glory, with multiple coloured rings. It is centred on the camera, which was obviously in front of the photographer's chest.

SKY AND CLOUD COLOURS

The normal daytime sky is blue because the oxygen and nitrogen molecules in the air strongly scatter the violet and blue wavelengths of the white light arriving from the Sun. Human eyes are insensitive to violet wavelengths, so the sky itself appears blue and the Sun, when high in the sky, appears yellow. When close to the horizon at sunrise or sunset, or when the light has to follow a long path through the atmosphere – such as beneath an extensive layer of clouds – only the yellow, orange and red wavelengths are able to reach our eyes.

The depth of colour to the blue sky is an indication of the air's purity. Water droplets scatter white light, so when they are present the air appears lighter in tint. The paler blue colours that occur with increasing distance (a phenemenon known as aerial perspective) are used subconsciously by the mind to gauge distance. At high altitudes, where the air is clear, the sky appears a deep, dark blue.

Solid particles suspended in the atmosphere also scatter light of all wavelengths, giving a lighter tint to the sky. Depending on their size, however, they may affect specific wavelengths. On very rare occasions, for example, the sizes may be such that they scatter away orange and red wavelengths, so that the Sun or Moon appears green or blue. Solid particles also tend to absorb light and thus give rise to a dark haze. This tends to build up during the day, and often results in a brownish layer close to the surface, which is most easily seen around sunset.

Slightly different effects arise when there have been major volcanic eruptions that have ejected dust (ash) or sulphur dioxide droplets high into the atmosphere. The dust layer is often illuminated by the rising or setting Sun, and has a distinctive appearance with striations running across the sky, roughly at right-angles to the direction of illumination. The sulphur dioxide droplets, by contrast, combine with water to create sulphuric acid droplets. These scatter red light from the Sun, which mixes with the normal blue of the sky to

▼ *Although the exact date of this photograph of the purple light is unknown, it was almost certainly taken during the period following the violent eruption of Mt Pinatubo in 1991.*

◄ *A sharp contrast between shadowed clouds, which appear almost black, and cumulus mediocris in the background, fully illuminated by the Sun.*

produce a twilight arch above the setting Sun (normally pale blue, with yellow, orange and red coloration towards the horizon) with a highly characteristic, vibrant purple hue. This effect, known as the purple light, is extremely striking visually, but notoriously difficult to photograph, because of the limitations of film emulsions and camera sensors. No strong occurrences of the purple light have been seen since the eruption of Mount Pinatubo in the Philippines in 1991.

Cloud colours

Cloud droplets are extremely small and efficiently scatter light of all wavelengths, which is why a cloud's basic colour is white. They are also present in such vast numbers in the majority of clouds that little light penetrates into the interior of the cloud. When a cloud starts to disperse, the smallest droplets evaporate first. The remaining larger droplets absorb a considerable amount of light, so old clouds may appear darker, especially when seen against the background of a younger cloud. For the same reason, clouds that have started to evaporate often show a darker edge around the main body of the cloud.

The colours of the clouds themselves are, of course, also affected by the light that reaches them from the sky and Sun. Clouds that are not illuminated directly by sunlight often appear bluish because they are reflecting blue light from the sky. Deep clouds generally have dark bases and this is particularly noticeable with large cumulonimbus clouds. This may sometimes enable the location of such clouds to be determined, even when the tops of the clouds themselves are hidden by (say) intervening stratiform clouds.

When the undersides of clouds are illuminated by orange and red wavelengths from the setting Sun to give a 'red sky at night', they are indeed an indication that the weather is likely to improve. Weather systems generally approach from the west, so red clouds indicate that the sky to the west is clear, and any system is moving away to the east. Conversely, a 'red sky at morning' suggests that the eastern sky is clear, but that clouds and changeable weather may be moving in from the west.

WEATHER SYSTEMS

There are various weather systems which have a recognizable structure and produce distinct sequences of weather. The most important of these, for anyone living in temperate latitudes, are depressions and anticyclones (low- and high-pressure systems, respectively), which extend their influence over large areas of the globe. Showers (which also include thunderstorms) are far more localized, but also show a distinct pattern of behaviour and the largest such systems may bring violent weather, including damaging hail and tornadoes. Far more extreme are tropical cyclones (also known as hurricanes and typhoons), which bring additional hazards.

The growth and decay of depressions

As mentioned earlier, the Polar Front in each hemisphere is a line of great contrast between cold polar air and warm tropical air. The winds on either side of the front are in opposite directions, running parallel to the isobars (a). Such a situation is not stable. The polar front may exist for a short while as a quasi-stationary front, but normally a wave is quick to develop (b). This gradually grows until a low-pressure centre forms, the air begins to flow across the isobars and a closed circulation develops (c). Distinct warm and cold fronts arise and there are significant changes in wind direction across these fronts. The whole system moves towards

▼ *The development and decay of a depression from a quasi-stationary front between polar and tropical air masses (a), an initial wave and closed circulation (b), maturity (c) and early decay (d).*

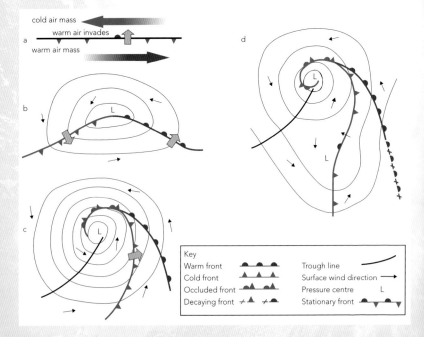

Key
Warm front
Cold front
Occluded front
Decaying front
Trough line
Surface wind direction
Pressure centre   L
Stationary front

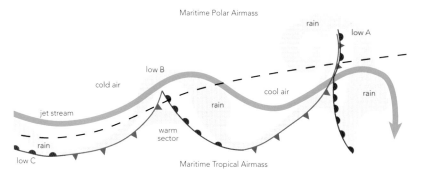

Maritime Polar Airmass

rain

low A

low B

cold air

cool air

rain

jet stream

rain

warm
sector

rain

low C

Maritime Tropical Airmass

▲ A family of
depressions and their
relationship to the jet
stream. The curvature
of the jet becomes
more extreme from the
youngest depression
(low C) to the oldest
(low A). Grey areas
indicate typical regions
of heaviest rainfall.

the east, partly guided (and possibly strengthened or weakened) by the jet stream at a higher level above it.

By the time a depression has reached an advanced stage, the cold front (i.e., the Polar Front) has often pushed down towards the tropics. A wave often develops to the rear of the initial depression and, in turn, becomes another fully fledged depression. There is a tendency for the two depressions to circulate around one another, i.e. (in the northern hemisphere), the first depression centre tends to move north and west, while the second centre moves south and east. A whole series of secondary depressions may be formed, one after the other, gradually advancing the cold front towards the equator. Generally, however, each succeeding depression is weaker than the one to the east, and the sequence eventually breaks off and subsequent depressions form closer to the pole.

The depression develops a distinct warm sector between the warm and cold fronts. This starts as a broad wedge, but the dense cold air behind the cold front causes the latter to move faster than, and thus overtake, the warm front to the east. The warm sector gradually becomes narrower, until eventually the cold front reaches the warm front, which it undercuts and lifts away from the surface, giving rise to what is known as an occluded front (d). The point at which this occurs is called the 'triple point', for obvious reasons. A pool of warm air is lifted away from the surface and lies above the occluded front.

Once this state has been reached, the depression starts to decay, although it may take a very long time to do so and may still produce extremely severe weather with strong winds and heavy precipitation. Occluded fronts are notorious for remaining stationary for long periods over a particular location and giving rise to heavy, continuous rain, which may lead to severe flooding. Eventually, however, the winds subside, the closed circulation ceases and the precipitation declines.

◀ *An occluded depression, covering much of the British Isles, with a well defined cold front stretching down into the Atlantic and a warm front over France, Germany and Italy.*

The diagram (p.67) shows how the jet stream lays above the depression. Its effects on the depression's development will not be described here, but note how it gradually develops into a more extreme 'S' shape as the depression ages.

An approaching depression

Jet-stream cirrus (p.21) is sometimes the earliest indication that a depression is approaching from the west. More commonly, high cirrus cloud appears and slowly increases and thickens until it covers the sky. Similarly, aircraft condensation trails (contrails) become persistent rather than dissipating and may blanket a large area of the sky. The increasing cirrus and persistent contrails are indications that the air aloft is becoming more humid ahead of the surface warm front. Generally, when the cirrus has thickened into a more or less continuous sheet of cirrostratus, various halo phenomena (p.54) are visible, but may persist for just a short time, depending on how quickly the cirrostratus itself thickens.

Another indication of an approaching warm front may be given by the wind direction at different altitudes. When the observer is facing the front (in the northern hemisphere), the low-level clouds tend to move from left to right, the medium-level clouds move straight towards the observer and the highest clouds (cirrus or even jet-stream cirrus) may even move from right to left. Such a 'crossed-wind' situation where the winds veer – i.e., change clockwise (p.15) – with increasing height is a sure sign that the approaching depression will carry the warm front, the warm sector and the subsequent cold front over the observer's position. (Conversely, if the winds at increasing altitudes back – i.e., change anticlockwise – then the weather is almost certainly likely to improve.) In the southern hemisphere the winds will initially back with height.

The speed at which the cloud changes from the first wisps of encroaching cirrus to thick cirrostratus gives an indication of the speed of the approaching system. If the change is rapid, taking just a couple of hours, the warm front and its accompanying rain are likely to arrive in just a few hours. If, in contrast, the change takes the best part of a day, it may be another day or two before the worst weather arrives.

The warm front

The slope of the warm front is very shallow, typically between 1:100 and 1:150 (1% and 0.67%). As we have seen earlier, the base of cirrus clouds is generally around 6 km (20,000 ft), so when we see the first signs of cirrus thickening into cirrostratus overhead, the surface warm front may well be 600–900 km (370–560 mi) away, normally to the west or south-west (in the northern hemisphere). An average speed of approach is about 50 kph (31 mph), which means that rain may be expected in about 9–10 hours and the warm front itself will arrive in 12–18 hours' time.

With a typical warm front at which warm air is rising, there is a distinct succession of cloud types as the front approaches. In the cool air ahead of the front, the low clouds will initially be ordinary cumulus and then flatten into cumulus humilis as the warm air starts to arrive overhead.

The principal sequence of clouds that accompany an approaching warm front is:

- cirrus
- cirrostratus (often with halo effects)
- altostratus
- nimbostratus

There is generally a transition from cirrostratus to altostratus and then to nimbostratus, but on occasions, particularly when well away from the low-pressure centre, some cirrocumulus and altocumulus may be present as the clouds thicken.

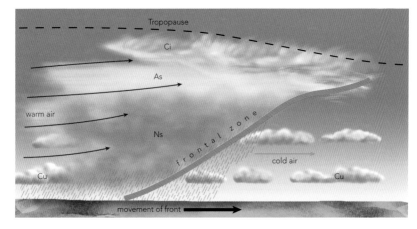

The sky gradually darkens as the clouds encroach. When the altostratus arrives, initially the Sun will appear as if through ground glass, but it will gradually disappear and objects on the ground will lose distinct shadows. There may be some slight precipitation from altostratus, but the arrival of the nimbostratus will herald the onset of the main rain-belt. This dark cloud, with its ragged base, is extremely thick and may reach down so far that it covers the tops of hills and tall buildings. The precipitation may cool the lower air to such an extent that ragged scraps of clouds (known as pannus) form within it below the main cloud.

Generally the main belt of more or less continuous rain will last for some three to four hours, but if the wind has dropped it may persist much longer. Bands of heavier rain lie ahead of, and roughly parallel to, the surface warm front, but there may be even heavier showers where there is instability and convection in the warm air aloft. In winter, freezing rain or snow may occur ahead of rain if temperatures in the cold air ahead of the front are sufficiently low and, naturally, under extreme conditions all of the precipitation may be in the form of snow.

While the changes in cloud cover occur, atmospheric pressure drops because the centre of the low is approaching the observer, but as the surface warm front arrives, the pressure steadies at its lowest point. The surface wind, which has tended to back slightly and strengthen more or less continuously, perhaps becoming gusty, now veers sharply with the passage of the front, swinging round from southerly to south-west or westerly. The continuous rain eases and peters out as the warmer surface air arrives and the cloud normally lightens and may start to break up. (In the southern hemisphere the wind initially veers slightly and then backs sharply as the front passes.)

▲ A warm front where the warm air is rising and sliding above the retreating cold air. The width of frontal zones varies widely, from as little as 10 km (6 mi) to 200 km (124 mi). Height is greatly exaggerated.

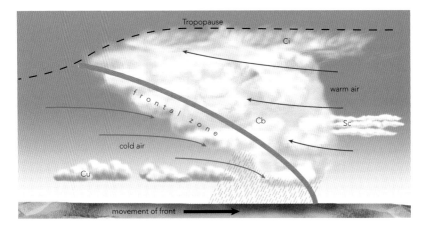

Tropopause

Ci

warm air

f r o n t a l z o n e

Cb

Sc

cold air

Cu

movement of front

▲ *A typical cold front found in depressions, where the cold air is undercutting the warm air. A similar structure is shown by isolated line squalls.*

It should be noted, however, that fronts are not as sharp a demarcation line as they may appear on charts, but generally consist of a zone of mixing between the two air masses. This zone has a typical thickness of about 1 km (3,300 ft), so at the surface it may be as much as 150 km (90-odd mi) wide, across which there is a transition from one air mass to the next.

The warm sector

The weather experienced in the warm sector depends entirely on how close one is to the low-pressure centre. If the centre is hundreds of kilometres away, the weather may be very pleasant with a warm breeze, vivid skies and well broken cloud at all levels, lit by bright sunshine. In summer, over the land, the heating may be sufficient for convection to occur and give rise to showers and even thunderstorms. If the parent low's centre is close, the rain may persist right across the warm sector and only cease once the cold front has passed. Similarly, the cloud may not break up completely with the arrival of the warm air. The cloud cover may remain in the form of stratus or stratocumulus.

The cold front

A cold front, which is generally faster moving and steeper than a warm front, undercuts the warm air, and has a typical slope of 1:50 to 1:75 (2% to 1.3%). In general, the approaching front is hidden by clouds in the warm sector, but occasionally there may be clear skies that allow the front to be seen. The rainbelt associated with a cold front is narrower and of shorter duration than that at a warm front, and is often the site of significant convection. It is quite common for there to be a belt of rain ahead of the surface front itself. The rain is generally very heavy and frequently accompanied by thunder. As the cold front passes, the wind veers again, perhaps from south-west or west to west or north-west; the temperature drops and the pressure rises. (In the southern hemisphere the wind backs as the front passes, becoming west or south-west.) In many cases there is a 'clear slot', relatively free of cloud, immediately behind the cold front and this is often easily visible on satellite images. The cold polar air behind the front brings good visibility and is often very unstable, especially after it has travelled over a warm sea. The instability produces numerous cumulonimbus clouds and associated showers. Over a continental region, such as the interior of North America, the air behind a cold front rarely experiences surface heating and therefore does not tend to give rise to showery conditions.

Occluded fronts

With an occluded front, there is no warm sector, because the cold front has caught up with the warm front and lifted a pool of warm air away from the surface. The warm and cold fronts are, in effect, combined, so the sequence of clouds associated with a warm front is followed immediately by the convective cloud found at a cold front. There are two forms of occluded front, which differ depending on which of the two air masses is the colder. In most cases, the following cold air mass (i.e., the one encroaching from the west) is colder, and tends to undercut the cool air to the east. In such a cold occlusion (below, left) the wind will (in the northern hemisphere) veer very sharply (by as much as 45°) and a pronounced drop in temperature will take place as the occlusion passes. Less often, the air to the east is colder and such a warm occlusion (below, right) tends to be less distinct, less extreme and have a much shorter lifetime.

Despite their seemingly simpler nature, occluded fronts may give rise to long periods of rain, especially when they become stalled. The persistent, nearly continuous rain that they produce often leads to severe flooding.

Away from the depression centre

Observers to the north of the depression centre (in the northern hemisphere) may note high cirrus overhead that

▼ *The structure of a cold occlusion (left) and a warm occlusion (right). In both cases a pool of warm air has been lifted away from the surface. Although looking innocuous, occluded fronts may give rise to extended periods of rain.*

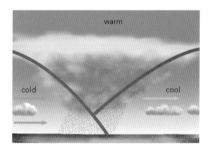

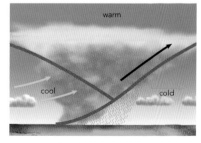

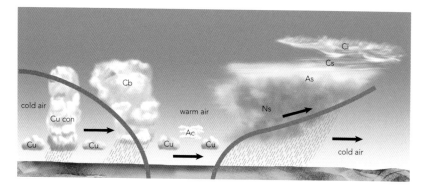

Labels on image: Ci, Cs, As, Cb, cold air, Cu con, warm air, Ns, Cu, Cu, Cu, Ac, Cu, cold air

gradually turns into cirrostratus. This may lead them to sus-pect that a warm front is approaching, particularly if the pres-sure begins to drop. Close examination of the surface and upper winds, however, will show that they are flowing in roughly parallel, but opposite, directions. The cloud does not build up significantly and later tends to disperse. The wind backs from south-east to east and finally to north-east and north as the depression centre passes to the south. As it does so the pressure rises.

Well to the south of a depression centre, the cirrus clouds may be succeeded by cirrocumulus and altocumulus, but these never completely cover the sky. Pressure and wind changes are slight and very slow, and high-pressure condi-tions soon return. In many cases, a ridge of high pressure may extend polewards behind the warm front and prevent the associated cold front from passing overhead.

▲ A schematic section across a depression passing through the warm sector. The clouds in the warm sector vary greatly and often include extensive stratocumulus and stratus.

Isolated warm and cold fronts
Isolated fronts may occur away from depressions, dividing air masses of differing temperatures and humidities. These are more frequent over continental interiors (such as over North America) than over maritime regions. They resemble the classic warm and cold fronts, although the warm fronts often display more marked convective activity and generally clearer skies. Isolated cold fronts are notable for a distinct drop in pressure before they arrive, followed by a sudden rise.

Systems where the warm air is subsiding
In the classic form of depression that has just been described the warm air is rising at both warm and cold fronts – which are known technically as 'anabatic fronts' or 'anafronts' (see box overleaf). In certain circumstances, how-ever, the warm air may be subsiding. The fronts are then known as 'katabatic fronts' or 'katafronts'. Such a system is much simpler, much weaker and less distinct. The subsiding

air suppresses convection, and does not give rise to the distinct sequence of high-level clouds that is such a useful sign of an approaching warm anafront. The cumulus ahead of the system tends to thicken into stratocumulus, which may become very heavy, but, as is generally the case with stratocumulus, does not produce much precipitation. What does fall is often in the form of light drizzle.

The temperature and winds change at the katafronts in a similar manner to the changes with the 'classic' frontal system, but less markedly and rather more slowly. Behind the warm front the stratocumulus thins and may allow the sky to be seen through small gaps, finally partially turning into stratus. At the cold front the stratus and stratocumulus thicken again, producing light rain. Once again, the temperature and wind changes at the front itself are slow and slight and the front is followed by cumulus or cumulonimbus clouds.

These low-pressure systems where the warm air is subsiding are therefore rather dull and uninteresting, but they are

▼ *A depression where the warm air is subsiding, producing a kata warm front (top) and a kata cold front (bottom). Compared with systems where the warm air is rising, both activity and clouds are greatly subdued.*

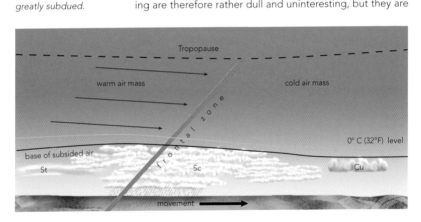

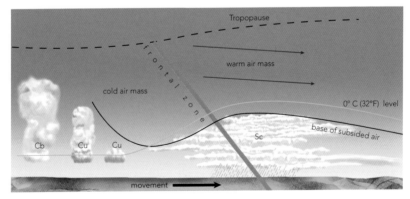

Anabatic and katabatic

The terms 'anabatic' and 'katabatic' (derived from Greek words) are used by meteorologists to describe ascending and descending air or winds, respectively. They are most commonly encountered in connection with anabatic and katabatic winds, otherwise known, respectively, as valley and mountain winds (p.94). The corresponding prefixes: 'ana-' and 'kata-', are used to describe frontal systems where the warm air is rising or descending.

very common in many areas of the world, especially in winter. With any frontal system, if you do not have access to forecast charts (p.153) it is not easy to judge the amount of rain that is likely to occur, but the main sign is that the skies become dull and dark, which indicates that the clouds are deep enough for the cloud droplets to grow into raindrops. If rain is forecast, the speed of the approaching high and medium clouds also gives an indication of whether it is likely to arrive during the day. If the clouds are moving very slowly, the rain may never materialize.

Anticyclones

As mentioned earlier (p.15) there are two forms of anticyclone (or high-pressure region): the shallow cold anticyclones, which typically form under winter conditions, and warm anticyclones, where air is descending throughout the troposphere. With cold anticyclones temperatures are very low and the sky may be completely clear. At other times an inversion may be present, which restricts the growth of any cumulus clouds so they tend to spread out as stratocumulus. Outside the polar regions, high-pressure areas tend to be in the form of a ridge, rather than a true anticyclone with a closed circulation around it. When skies are clear, temperatures generally drop extremely rapidly at night.

In a warm anticyclone, the descending, warming air tends to suppress any cloud formation, but there may be some scattered cumulus clouds or well broken stratocumulus and – if the surface air is part of a warm, moist, maritime tropical air mass – extensive fog or stratus cloud may occur, especially with night-time cooling. The cloud or fog may be very persistent, especially in winter if winds are light, but in summer, heating during the daytime usually disperses it. Overcast skies with thick stratus or stratocumulus cloud may sometimes linger for many days (or even weeks), giving rise to what is known as 'anticyclonic gloom'.

In a warm high, air descending throughout the troposphere acts as a barrier to the westerly winds and to the normal eastward motion of depressions. Such 'blocking highs' may be very stable and bring unseasonal weather to the areas around them. The Polar Front lies closer to the pole than normal, and depressions are steered north or south of their usual tracks. A winter blocking high over Scandinavia, for example, may bring bitter easterly winds to western Europe and force depressions to track over the Iberian Peninsula and the Mediterranean, farther south than normal.

◄ *An anticyclonic situation over Central Europe, with the Polar Front across northern France, the Netherlands and northern Germany. It is preventing depressions from travelling east, forcing them to travel over the British Isles, Norway and Sweden.*

Showers

To meteorologists, 'showers' are relatively localized precipitation from either cumulus congestus or cumulonimbus clouds. As mentioned earlier (p.32), in temperate regions rainfall from cumulus congestus occurs primarily in summer, but the predominant sources of non-frontal precipitation at all times of the year are cumulonimbus clouds. These grow from cumulus clouds until freezing occurs at the highest levels, with the falling ice particles melting at a lower level to give rain or, if temperatures remain low enough, snow. Energetic cumulonimbus clouds may produce hail or develop into thunderstorms.

Cumulus and cumulonimbus grow only when there is instability, with higher temperatures near the surface than at greater altitudes. The overall amount of convection is revealed by the area covered by clouds, while areas of clear sky indicate the amount of sinking air that balances the rising thermals. So as a rough guide, showers are likely when the area of cloud exceeds the area of clear blue sky. Local features, such as towns and cities, may provide 'hot spots' that assist convection and air rising over hills or mountains frequently initiates strong uplift and cloud growth. Clouds that are very deep also have very dark bases, which is another indication that they are likely to produce showers.

Again as mentioned earlier, individual cumulonimbus cells have a limited lifetime consisting of roughly equal periods of growth, maturity and decay, with an overall duration of one to one-and-a-half hours. Such small clouds may cover an area of 10–12 sq km (4–5 sq mi) and the rainfall may last between 10 and 30 minutes. Over land, the growth and duration of suitable convective conditions is largely dependent on daytime heating and showers tend to die out at night as convection ceases in the late afternoon and early evening. When convection arises from cool air passing over a warm sea, activity may continue throughout the night.

When the gradient wind (wind blowing parallel to curved isobars) is strong, showers tend to be brief, lasting less than half an hour and with moderately long periods of sunshine between them. When the wind is weak, the showers may last longer, but there are fewer of them during the day. Showers themselves create blustery winds as a result of the strong updraughts and downdraughts that occur within cumulonimbus clouds. For example, the updraughts tend to draw air in from the area ahead of the cloud, producing a localized surface wind blowing towards the cloud. This may confuse inexperienced observers, and is the cause of the statement often heard that a shower approached 'against the wind'. The changes in strength and direction may be considerable:

▼ *An individual cumulonimbus cloud producing a very heavy shower of rain and possibly hail. As with most showers, only a limited area of the ground is affected.*

the surface wind may easily double in strength and reverse (or change) direction a number of times as a shower passes overhead. A shower also causes a drop in the surface temperature, typically of about 3°C (5°F), because the rain tends to evaporate in the warmer air below it, creating a cooling effect.

The air drawn into vigorous cumulonimbus clouds frequently leads to the formation of additional convective cells, and there is often a cluster of distinct cells at various stages of growth, maturity and decay. The additional cells naturally prolong the activity of the overall cluster, but will normally bring rainfall to slightly different areas of the ground, depending on the prevailing gradient wind.

Hail

Hail consists of solid particles of ice more than 5 mm (0.2 in) in diameter (smaller particles are known as ice pellets or small hail). Hail is created as ice particles circulate through various regions of the cloud. Where supercooled water droplets are present, they freeze instantaneously on contact with an ice particle, trapping bubbles of air between them, and thus give rise to an opaque layer around the central nucleus. Where liquid water droplets are present, they tend to spread into a liquid layer before freezing, when they become a layer of clear ice. The powerful updraughts pres-

▼ *A hailstorm over New Mexico, USA. The distinctive white signature of the falling hail is clearly visible.*

◀ *Typical medium-sized hail, clearly showing that the stones have been built up from layers of clear and opaque ice.*

ent in vigorous cumulonimbus clouds carry the hailstones back towards the top of the cloud, only for them to descend again. Hailstones therefore grow larger and larger, usually with alternating layers of clear and opaque ice, until they become too heavy to be supported by the updraughts, falling from the base of the cloud at speeds as high as 150–160 kph (93–100 mph). Even small hail may cause extensive damage to crops and other objects on the ground, and some stones more than 200 mm (8 in) across have been recorded. The largest known individual hailstone fell at Gopalganj in Bangladesh on 14 August 1986 and weighed 1 kg (2.2 lb). Clumps of hailstones frozen together have fallen, weighing as much as 4 kg (8.8 lb).

Thunderstorms

Thunderstorms occur when the vigorous updraughts within cumulonimbus clouds separate positive and negative charges and sweep them into different regions of the cloud. The exact mechanism by which the charges become separated in the first place is still somewhat uncertain, but it appears to be related to the freezing and fragmentation of ice particles in the upper regions of the cloud. In general, the lighter, positively charged particles are lifted to the top of the cloud and the heavier, negatively charged particles accumulate at the base. As the cloud travels across the surface it induces a positive charge on the ground beneath it. The charge follows the cloud until eventually, either the charges become so great, or the distance between the cloud and the ground decreases to such an extent – usually over a tall object such as a building or tall tree – that the electrical resistance of the air breaks down and a lightning strike occurs.

A lightning strike is extremely hot – hotter than the surface of the Sun – and by heating the channel of air, causes it to expand at supersonic speeds, giving rise to the familiar crack

Lightning safety

Outdoors:
- Avoid open spaces where you are the highest object
- Take shelter in a building, under a bridge or in a car
- Do not stand under isolated trees
- On hills or mountains, move to lower ground if possible
- If there is no cover, choose the lowest possible spot, but avoid wet ditches or marshy areas
- Do not lie flat, instead crouch down on the balls of your feet, with arms around your legs and your head on your knees
- If your hair or that of companions begins to stand on end, drop to a crouch immediately
- At sea or on a lake, make for shelter as soon as possible
- Large yachts generally have lightning protection, but dinghies are liable to be struck, so ideally, wrap a length of (anchor) chain or wire rope around the wire shrouds and dangle the end in the water; avoid touching this makeshift conductor

Indoors:
- Stay away from water pipes, telephone and electrical wiring and aerial feeds
- Disconnect telephones, computer modems, etc.
- Disconnect vulnerable electrical appliances (computers, television sets, etc.) from the electrical supply
- If observing from within a room, avoid getting close to a window or leaning outside
- Observing from a veranda or porch is generally safe

of thunder. Light from the strike reaches the observer almost immediately, but the sound waves travel at a much slower rate. Counting the number of seconds between flash and thunder gives an estimate of the distance: 1 km for every 3 seconds difference (about 5 seconds per mi). If you note the position of the flashes and find that the bearing remains constant, this suggests that the cloud is likely to pass over-head. However, as with rainfall from cumulonimbus clouds, the duration of lightning from any one cell is limited, so if you have been watching a lightning cell for 20–30 minutes it is likely to die out soon, possibly before it reaches you. If no sound is heard, then the cloud is probably at least 25–30 km (15–18 mi) away.

Lightning discharges occur not only between cloud and ground, but also between different regions of a cloud and between separate clouds. When the discharge channel is visible, people often describe it as 'fork lightning', in con-trast to 'sheet lightning', where the channel is invisible.

There is no real difference. With discharges within a cloud, or from cloud to cloud, the channel may be hidden by the intervening cloud giving the more diffuse appearance. With cloud-to-ground lightning the branching nature of the discharge channel is more clearly visible, particularly in regions of the world where the cloudbase is high. 'Heat lightning' is merely distant lightning, seen on the horizon, that is too far away for any sound to be heard.

▲ *This photograph of a lightning strike over Dallas-Fort Worth Airport clearly shows the branching structure of 'fork' lightning, where only one channel reaches Earth and carries the main (upward) stroke.*

Multicell storms

As described earlier in connection with cumulonimbus clouds, a vigorous thunderstorm cell may draw the surrounding air into it and give rise to a new, active cell. It is often possible to see the line of growing cells trailing behind the active storm centre cells. Such a 'flanking line' generally lies on the southern rear flank, i.e., to the south or south-west of the main centre of activity (in the northern hemisphere).

When two or more cells are active, such systems are known as multicell storms and, simply because of their vigorous activity, they tend to persist for several hours, although the lifetime of each individual cell remains limited. It is usually possible to determine the location of each active cell and how likely it is to pass overhead. Even if you escape an active

cell that passes to the north of your location (in the northern hemisphere) you may well be affected by a slightly newer cell that lies to its south.

Supercell storms

The most extreme individual thunderstorms are those that are known as supercell storms, which arise when there is a deep pool of unstable air and winds increase with altitude. Here, instead of there being a number of separate cells, the system becomes organized into a giant rotating column of rising air, known as a mesocyclone, which extends high into the top of the cumulonimbus cloud, which may be anywhere between 8 and 15 km (5 and 9 mi) high. Such systems are accompanied by a complex set of updraughts and down-draughts, with cool air entering the system at middle levels. Because the main updraught bringing warm humid air into the system is widely separated from the downdraughts, a supercell system has a very long lifetime of six hours or more.

▲ *A multicell thunderstorm, with the closest cell on the right and one slightly farther away on the left. There are also suggestions of at least one, or possibly two other active cells on the far left.*

▶ *An advancing supercell over the Midwestern USA. The rings of cloud give some idea of how the circulation is organized into a gigantic, rotating column of air.*

Supercells are the most violent storms and may be accompanied by torrential rain, frequent lightning strikes and damaging hail. The rainfall may be so intense that it produces flash floods, and the downdraughts themselves may be so strong that they produce considerable damage. Supercells also spawn the most violent tornadoes. They are most frequent at middle latitudes, particularly over the central and eastern United States and other continental areas of the world.

Squall lines and mesoscale convective complexes

Occasionally, cumulonimbus clouds become organized into a vast cluster or line of instability, known as a mesoscale convective complex (MCC). As seen (and defined – in metric units) from satellite images, these cover vast areas with cloud (over 100,000 sq km below −32°C and 50,000 sq km below −52°C), and last more than six hours. (These figures approximately equate to 39,000 sq mi below 26°F and 19,500 sq mi below −62°F.) MCC systems tend to grow to maximum size overnight as smaller storm clusters merge and often persist well into the next day. They are most common over mid-latitude and tropical areas, particularly over the Americas, Africa and Asia. MCC systems that travel from the land over the ocean often act as seeds for the formation of tropical cyclones (hurricanes).

The linear form is known as a known as a squall line, which may be hundreds of kilometres long and which sweeps across country. (Similar, but normally less vigorous squall lines may form part of a fast-moving cold front.) Such squall lines may be so vigorous that they are accompanied by a slight rise in pressure where precipitation is greatest and a slight drop in pressure behind the storm.

The occurrence of thunderstorms

Thunderstorms are often associated with cold fronts, where they are concentrated in bands some 25–80 km (15–50 mi) wide and extending for hundreds of kilometres. They do sometimes occur on warm fronts, but such activity tends to be sporadic and short lived, with perhaps no more than one or two flashes of lightning. In maritime areas, thunderstorms are often found in the cold air mass that follows the cold front. Over continental areas, thunderstorms tend to be most frequent in the summer when there is intense heating of the land. Supercell storms are most likely to be encountered over continental areas in summer, or at any time of year in the tropics.

Individual circumstances may, however, alter the likelihood of major thunderstorms. In western Europe, for example, during summer, a layer of potentially unstable hot air, known as the Spanish Plume, may build up over the Iberian Peninsula and spread north over France. It acts as an inversion, subduing convection beneath it where the air becomes extremely humid. If a low-pressure trough in the upper air passes over the area, convection breaks through the inversion and produces violent multicell or supercell storms. Such storms sometimes cross the English Channel and affect southern England. A similar phenomenon, known as the Mexican Plume, affects the southern states of the USA.

The tops of storms forming in polar air may often be seen from distances of 80 km (50 mi) as they approach, when they bring squally winds. Storms that move up from the tropics are even higher, so their anvil tops may be seen from even greater distances. The weather in advance of the storm is often distinctive, as the day becomes heavy, humid and sticky and the sky becomes overcast and very dark, occasionally lit by the approaching lightning.

Whirls, devils, spouts and tornadoes

There is a whole family of phenomena, known collectively as whirls, all of which are relatively small, concentrated rotating columns of air, of varying intensity. The weakest of these are the whirls or devils, which are named for the material that they raise from the surface. Some of these, such as dust devils or sand devils, arise when there is intense heating of the surface, often accompanied by differences in friction with the ground, which initiates the rotation of the column of air. Other devils may occur when air is funnelled into an area by nearby hills. Water and snow devils often arise in this way. All devils tend to be short lived, persisting for a few minutes or tens of minutes.

Rather stronger are the whirls known as waterspouts and the corresponding landspouts – a term increasingly used in recent years. These arise from strong updraughts and downdraughts created by vigorous cumulus congestus or cumulo-

▶ *A typical dust devil, seen in Kenya. The tops of such devils normally lie in clear air, free from cloud.*

nimbus clouds and usually begin as a cone of cloud, known as a tuba, that descends from the cloudbase. When this reaches the surface it raises a 'bush' of water or dust and, in the case of a waterspout, creates a 'dark spot' on the surface of the water (overleaf). Both landspouts and waterspouts may be strong enough to cause some damage, although there are some suggestions that waterspouts are slightly less dangerous. Some tornadoes – perhaps better described as 'tornadic vortices' – that are spawned by tropical cyclones and (occasionally) by individual vigorous cumulonimbus clouds resemble this class of whirls.

The most destructive whirls are tornadoes, popularly known as 'twisters'. True tornadoes, which may be highly destructive, are created by supercell storm systems, which as previously mentioned contain a vast, rotating cylinder of air (a mesocyclone). The appearance of a tornado is often preceded by a lowering of the base of the cloud, in an area known as the rain-free base, giving rise to a rotating wall cloud. A tornado may then descend either from the wall cloud or the rain-free base. The extreme pressure drop of at least 200–250 mb within the funnel cloud causes the air to become saturated and condense, producing the opaque grey column.

The violent motions within a supercell may give rise to extreme wind speeds in the associated tornadoes. (The highest measured speed is 512 kph (318 mph) for a tornado on the outskirts of Oklahoma City in 1999.) Upcurrents of more than 250 kph (155 mph) have also been recorded, and these are capable of lifting even extremely heavy objects (such as locomotives) from the surface. Sometimes a column lifts away from the surface and touches down again in a different place. Typically the diameter of a tornado at the surface is 100–200 m (330–660 ft), but monster tornadoes as wide as 2 km (6,500 ft) have been recorded. Such giant tornadoes are sometimes accompanied by subsidiary, smaller vortices arranged around the outer edge of the main column.

Tornadoes often occur in swarms – as many as 100 in a single day – as their parent storm systems move across country. They are often erratic in their duration and motion. Some may remain stationary for as long as 45 minutes, while others decay after just a few minutes. Some have been found to persist for hours and cover tracks as long as 400–500 km (250–300 mi).

▲ *A waterspout photographed over the Mediterranean Sea. The 'bush' of spray at the base of the spout is just visible.*

Tropical cyclones

The violent rotating storms known technically as tropical cyclones have different names in different parts of the world:

- hurricane – Atlantic and eastern Pacific Oceans
- typhoon – north-western Pacific Ocean
- cyclone – Indian and south-western Pacific Oceans

By definition, a system becomes a tropical cyclone when wind speeds exceed 33 m/s (approximately 120 kph (75 mph)). All consist of a low-pressure core surrounded by spiralling bands of intense convection. Satellite images show an immense cirrus shield above each system, where air is flowing out at altitude. Such cirrus shields may be thousands of kilometres across.

Tropical cyclones are able to develop only over oceanic waters that exceed 27°C (81°F) at the surface. (Just one

▶ *(Top) An incipient tornado: a funnel cloud (tuba) reaching down towards the surface. This photograph clearly shows the rotating wall cloud surrounding the descending funnel cloud. (Bottom) A mature, highly destructive tornado, clearly showing the debris cloud raised from the surface.*

◄ *The notorious hurricane, Katrina, over the Gulf of Mexico in August 2005, shortly before it made landfall and devastated New Orleans.*

system is known from the South Atlantic, where sea temperatures are normally too low.) Additional requirements for their formation include low wind shear (i.e., little change in speed or direction) with increasing altitude and divergence aloft (which draws air upwards from the surface). In general, systems develop 5–10° north or south of the equatorial low-pressure zone. Systems decay when they travel over colder waters or pass over land, where their sources of energy die away. They may then sometimes merge with existing weather systems to produce strong extratropical cyclones (i.e., depressions).

The low-pressure core, the eye, where warm air descends in the centre of the storm, may be cloud-free. It is surrounded by the eyewall, the site of the most vigorous convection, the highest cumulonimbus cloud bands, the most intense precipitation and the most violent winds. The devastation produced by tropical cyclones arises from three factors. First, the extreme winds, which may be accompanied by even stronger subsidiary tornadoes, are usually responsible for the most immediate damage. Second, the low pressure in the centre and the high winds act to create a dome of water beneath the storm. As this approaches the shore it creates a high storm surge, which may be as much as 10–12 m (33–39 ft) above normal water levels. Such storm surges may travel 10–15 km (6–9 mi) inland and bring great destruction and loss of life. Finally, the torrential rains may cause widespread flooding, with mudslides and landslides, which may them-selves produce many casualties. However, it should be noted that much tropical agriculture is dependent on the rainfall from tropical storms.

▼ Flooding caused by Hurricane Katrina's storm surge in New Orleans. Over 300,000 people are thought to have been killed by a storm surge in Bangladesh in 1970.

LOCAL WEATHER

Although major weather systems affect areas that cover thousands of square kilometres, there are many local effects that create significant variations in the weather experienced at any particular location. The local topography – especially valleys, hills and mountains – exerts a strong influence, as does proximity to the sea or other large bodies of water.

Friction and turbulence

As we have seen earlier, the Earth's surface affects the wind through friction. Obviously this effect occurs least over the sea, but may increase greatly over the land, and is the main reason why a strong wind on the coast may hardly be noticed some distance inland. The friction generates turbulence in the boundary layer (p.15) and this turbulence means that this layer is thoroughly mixed and therefore relatively uniform in its properties, such as temperature and humidity. Turbulence caused by a rising wind during the day often leads to the dispersal of fog and mist that have formed overnight. Turbulence is often extreme in built-up areas, where the buildings may cause considerable gustiness and changes in wind direction.

Air at higher levels flows more freely, but on meeting an obstacle such as a hill or mountain is forced to rise and accelerate, spilling over the crests and rushing down the leeward slopes. With weak winds this may result in little more than increased turbulence, but with stronger winds even relatively low hills may generate major eddies on the leeward side and a train of waves that extend downwind. When the atmosphere is stable, wave (or lenticular) clouds – described earlier (p.50) – often form in the crests of these otherwise invisible waves and may persist for a long time. Even when no clouds form, the waves and turbulence are still present and they are often responsible for breaking up a layer cloud (such as stratocumulus) leeward of the hills,

▼ *A schematic representation of the formation of wave clouds, here shown as if there were three moist layers. Sometimes the waves may be so strong that a rotor forms downwind of the mountains.*

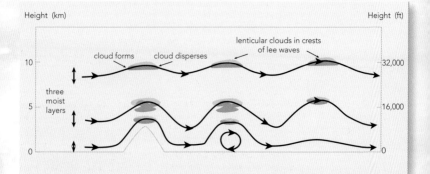

Height (km)
Height (ft)

cloud forms   cloud disperses

lenticular clouds in crests of lee waves

10 — 32,000

three moist layers

5 — 16,000

0 — 0

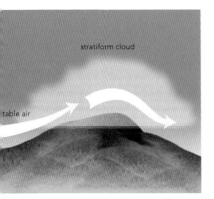

▲ Under stable conditions (left), a cap of cloud may form over the high ground, but disperse downwind. When conditions are unstable (right), the orographic lifting may be sufficient to trigger the formation of major cumuliform clouds.

even though elsewhere over flat country the same type of cloud may remain unbroken. Under certain conditions, particularly when high mountains are windward of a deep, broad valley, a giant eddy, known as a rotor, may form over the valley, where the surface wind is opposite to the gradient wind above.

Hills and mountains: clouds and rain

Orographic uplift (p.24) over hills or mountains obviously has a great effect on cloud cover and precipitation. In general, if uplift generates cloud over the hills, the cloudbase will be lower on the windward side than on the leeward where the air is descending. Depending on circumstances, cloud may shroud the hill tops or there may be clear air below the cloudbase. The exact form of cloud will, naturally, depend on whether conditions are stable or unstable. With stable conditions, stratiform cloud forms, and this tends to disperse leeward of the hills. Similarly, cumuliform clouds form under unstable conditions and again may occur only over the hills or mountains. If an approaching frontal system is bringing warm, humid air that may be expected to rise over the hills, any walkers or climbers would do well to be cautious, because there is the strong likelihood of limited visibility and rain or even snow may be produced. Although stratiform cloud may not produce very much precipitation, anyone unlucky enough to be enveloped by the cloud will still become very wet.

When the range of hills or mountains extends across the wind direction much of the precipitation will fall on the windward slopes. To the leeward the land will be sheltered and dry, and this rain shadow effect is extremely important to the climate in many parts of the world where high mountains extend across the path of the prevailing winds. At high latitudes, in particular, the actual slope of the ground may also become important. Slopes that face away from the Sun (e.g., northwards in the northern hemisphere) may receive little or no winter sunshine. Slopes that face the Sun may receive a considerable amount of solar heating, especially in summer. But heating at any time of the year may cause strong thermal currents that may initiate development of cumulus clouds over the peaks and may lead to the rapid growth of cumulonimbus clouds, with their associated precipitation. Depending upon the exact circumstances such clouds may remain essentially stationary over the hills or mountains and may, in some cases, be the cause of dangerous flash floods.

With unstable air, even if no showers are being produced over lower land, even a slight ridge of high ground may provide enough extra lift to create showers. When showers are expected to be frequent, the additional orographic uplift may convert a shower into a 'cloudburst' or even a thunderstorm. In winter, the freezing level is closer to sea level than during the summer and even over relatively low hills may well be below the peaks, which thus become covered with snow, whereas lower down, the ice crystals melt and turn to rain. The snow line (the height above which snow may be expected to fall) varies throughout the winter, but does not lie below the level where the temperatures are above 3°C (37°F).

When air descends to the leeward of any high ground, it warms (at the saturated or dry adiabatic rates, pp.24–6). If it has deposited any of its moisture on the windward slopes or over the peaks, it will be warmer and drier at any level than at the same level to the windward (opposite). This effect (the föhn effect) may also occur if wave motions force air down to a lower level on the leeward side of high land. The föhn effect may occur to a greater or lesser amount under many conditions, but extremes occur near high mountains. The Föhn itself (after which the phenomenon is named) occurs over the Alps, often occurring when a depression to the north draws air from the south over the mountain barrier. In North America, the Chinook occurs on the eastern side of the Rockies and can cause extremely rapid temperature rises. (The record is 27 deg C (49 deg F) in two minutes at Spearfish, North Dakota, on 23 January 1943.) Such sudden changes often rapidly reduce snow cover, may trigger avalanches and (by drying vegetation or wood used in buildings) bring a substantial fire risk.

Hills and valleys

Valleys naturally have completely different effects on local weather than hills and mountains, although the combined differences in relief are also very important. Where there are gaps in a range of high ground, the wind tends to funnel through them and may become extremely strong. This is particularly the case when the valley and the isobars are aligned, and this is one of the causes of the strength of the famous

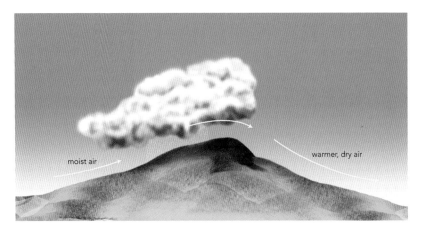

moist air

warmer, dry air

▲ *When orographic lifting causes precipitation on the windward side of hills or mountains, the air becomes drier and warms at a faster rate when it descends to leeward, giving rise to a hot föhn wind.*

French wind, the Mistral, that sweeps down the valley of the Rhône and out over the Golfe du Lion in the Mediterranean.

On clear nights, the sides of valleys radiate heat to space and cool the air in contact with them. The cold air slides downhill and accumulates in the valley bottoms, where the air is often humid, thanks to the presence of streams, rivers or even lakes. Cool, moist air readily forms mist, which often thickens into fog during the night. Similarly, valleys often accumulate freezing air and become frost pockets. When the general night-time temperature is near 0°C (32°F), the air temperature in valleys is often below freezing and the ground temperature even lower. The frost is hardest and mist or fog thickest around dawn.

By day, the situation is reversed. The ground on the valley sides rapidly absorbs heat from the Sun and warms the air in contact with it which then starts to rise. Any mist or fog starts to lift into low stratus, which begins to evaporate and disperse.

During the afternoon the heat concentrated in a valley may make it one of the hottest places in the area, and this is particularly true if the valley sides are much higher than the floor, as is often the case in mountainous areas. Whereas a lowland site might expect a temperature of (say) 20°C (68°F), in a valley the temperature might well be 10 deg C (18 deg F) higher. The high temperatures often produce strong thermal currents and heavy cloud over the intervening ridges and peaks, and even lead to significant precipitation. It is for this reason that hill-walkers and climbers are advised to start their climbs in the early morning so that they may reach the summit and descend again before it is shrouded in thick cloud. When the orientation of the valley is such that one side is strongly heated by the Sun, but the other remains largely in shadow, a very strong cross-valley circulation may build up, with similar results.

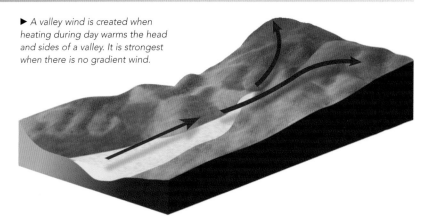

▶ A valley wind is created when heating during day warms the head and sides of a valley. It is strongest when there is no gradient wind.

Valley wind

Strong heating of a valley and its slopes during the day may be sufficient to cause a wind to blow up the valley and fan out towards its head. Such winds are normally strongest under otherwise calm conditions when there is little gradient wind, although if the latter is in precisely the same direction as the valley it may slightly intensify the flow. Such valley winds are also known as anabatic winds (p.75). They are generally weaker than the corresponding mountain winds (described next).

Mountain wind

If the temperature differences at night and the slope of the ground are sufficiently great, cold air will not simply accumulate at the bottom of a valley, but under the influence of gravity will flow down the centre of the valley towards lower ground. This wind is commonly called a mountain wind – technically known as a katabatic wind (p.75). Katabatic, or fall winds, are also produced by extensive cooling of air above a high plateau or an ice- or snow-covered area, often triggered by the motion of a pressure system, which causes the cold air to cascade down the slopes. Despite warming adiabatically, these fall winds may remain far colder than the lower air that they displace. Where the winds funnel through gaps between peaks, the winds may become very strong, and the extreme examples of this are where katabatic winds descend from the ice caps of Greenland and Antarctica. In the latter case, at Commonwealth Bay on the George V Coast, average katabatic wind speeds throughout the year are 67 kph (42 mph) and a peak of about 320 kph (200 mph) has been recorded.

Mountain and valley winds

There is often confusion – even among amateur and professional meteorologists – over which wind is which. It is sim-

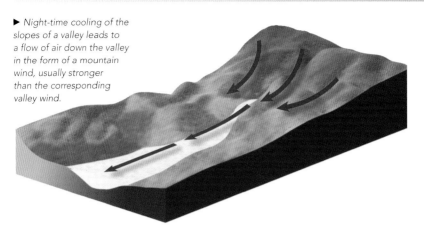

▶ *Night-time cooling of the slopes of a valley leads to a flow of air down the valley in the form of a mountain wind, usually stronger than the corresponding valley wind.*

plest to remember that winds are always named for the direction FROM which they come. So just as one has easterlies from the east, there are mountain winds from the mountains and valley winds from the valleys.

Coastal areas

Areas near the sea enjoy an important form of microclimate. The sea temperature changes relatively little over the year, and also very slowly, because of the high thermal capacity of the water. There is therefore a strong tendency for coastal areas to experience less extreme temperatures throughout the year, and they may be completely frost- and snow-free in winter, unlike areas farther inland. In winter, the relatively warm seas may produce weak damp thermals that rise just a short distance into the cold air before they become saturated, producing low sheets of cloud that are often carried inland by onshore winds, where they may cause gloomy

▼ *The cold air that accumulates over snow- or ice-fields cascades downhill as a katabatic (fall) wind. Despite warming as it descends, it usually remains very cold and strong even when it reaches the lowlands.*

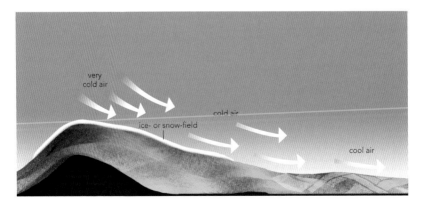

very cold air

cold air

ice- or snow-field

cool air

weather, with hill fog and drizzle. Somewhat similar cloud may form if warm air rises over sea cliffs or neighbouring hills, although such cloud is usually confined to the immediate vicinity of the coast and often dissipates later in the day when temperatures rise. In autumn and winter when the sea is relatively warm, cumulus cloud may arise over the sea, but not inland. If there is a strong contrast in temperatures, however, perhaps if the air is of polar origin, large cumulonimbus may build up, producing showers or even thunderstorms that move inland. For this reason, some regions, such as Scotland, experience more thunderstorms in winter than in summer.

The sea is normally still very cold in spring and therefore it tends to cool the air that comes into contact with it. When there is little or no wind, the air temperature may drop to the dewpoint, creating a sea fog. This often invades coastal areas during the day, although a weak sea breeze and turbulence over the land tend to cause it to thin. At night, when the wind drops, the fog thickens and often extends well inland.

Quite apart from any effects of cliffs or other high land, winds over coastal regions also tend to differ from the gradient wind that is found over inland areas. A westerly wind flowing along a coastline lying east-west, for example, experiences more friction over the land, reducing the wind speed, which (as explained earlier, p.14) causes the wind to back towards the south-west and become an onshore wind. Conversely, a wind off the land will tend to speed up over the water and veer to the right. These, and other effects, are obviously of great importance to sailors and all those using inshore waters.

Sea and land breezes

During summer, the land heats up rapidly during the day. Air inland becomes buoyant, less dense and of lower pressure than the air over the sea. The pressure difference causes low-level air to move inland in the form of a cool sea breeze, reducing convection along a coastal strip, which is typically cloud free. Farther inland, the boundary of the cool air forms a sea-breeze front, which, especially if it encounters higher ground, may be marked by convective cloud and even produce showers. Such sea-breeze fronts may penetrate many kilometres inland and be recognizable even late in the day. When sea breezes move inland from both sides of a peninsula, they may meet and combine to give exceptionally strong convection and precipitation, which has often caused extreme flash flooding in the affected area along the spine of the peninsula.

If the air is very stable (perhaps of maritime tropical origin), convection may not set in and no sea breeze will develop. Similarly, if the prevailing wind is offshore during the morn-

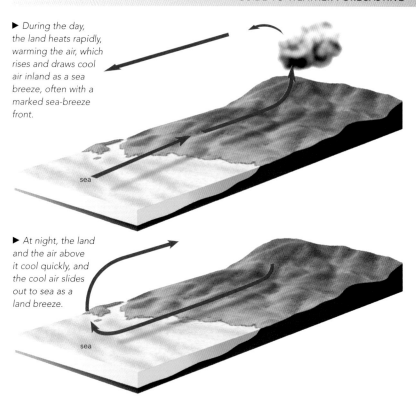

▶ During the day, the land heats rapidly, warming the air, which rises and draws cool air inland as a sea breeze, often with a marked sea-breeze front.

▶ At night, the land and the air above it cool quickly, and the cool air slides out to sea as a land breeze.

ing, the sea-breeze effect may not become sufficiently strong to overcome the gradient wind.

At night, the process is reversed. The land cools rapidly through radiation to space – particularly if the sky is clear – and air begins to flow towards the sea as a mild land breeze. A land-breeze front travels out to sea but, naturally, there is no orographic uplift to increase cloud as there is with a sea breeze. The land-breeze front may often be seen on early-morning satellite images, running roughly parallel to the coastline, and sometimes far out to sea.

Lake winds

Large bodies of water other than the sea may also give rise to similar effects, especially in the case of lakes that are surrounded by high ground. A lot depends upon the orientation of the valley and its exposure to the Sun. Daylight heating of the surrounding area may produce a lake wind, blowing away from the water, that closely resembles a sea breeze. This may be particularly strong if there are steep slopes that receive a large amount of radiation from the Sun.

At night, the area will often have a normal mountain wind. Very large bodies of water, such as the Great Lakes in North America and the Caspian and Black Seas in Eurasia, regularly produce strong lake winds.

Water and ice in the atmosphere

Technically, any form of water or ice that falls to the ground is known as precipitation. The most common forms are:

drizzle	drops below 0.5 mm (0.02 in)
rain	drops above 0.5 mm (0.02 in)
snow	clusters of ice crystals (p.99)
hail	solid lumps of ice (p.78)

Additional, less common, forms are:

diamond dust	tiny ice crystals created under very cold conditions
freezing rain	rain that freezes on impact (p.105)
ice pellets	frozen raindrops below 5 mm (0.2 in)
sleet	melting snowflakes or (in North America) ice pellets
small hail	snow pellets with thin ice coating
snow grains	flattened, opaque ice grains below 1 mm (0.04 in)
snow pellets	opaque ice grains, 2–5 mm (0.08–0.2 in)

▼ A photograph from an early Gemini space mission, showing a land-breeze front on both sides of southern India.

The definition of precipitation specifically excludes mist and fog, dew (p.102), frost (p.103) and rime (p.104), and also the particles or droplets in virga (p.43) that evaporate before they reach the ground.

The formation of rain by the coalescence and freezing processes (giving 'warm' and 'cold' rain, respectively) has been described earlier (p.82). The largest raindrops are about 2.5 mm (0.01 in) in diameter. Any drops that do happen to grow larger than this tend to split into two smaller droplets.

Certain cloud types, most notably stratus and stratocumulus, are normally able to produce precipitation only in the form of drizzle, although very thick stratocumulus may occasionally give rise to rain. Most rain falls from nimbostratus, cumulonimbus or cumulus congestus clouds, although it may also sometimes be produced by altocumulus (particularly altocumulus floccus, p.44) and altostratus.

Snow

When ice crystals reach the ground without melting they create a layer of snow. The ice crystals themselves are in the form of the beautiful, complex, six-sided shapes that are commonly described as 'snowflakes'. In fact, such crystals are extremely small and true snowflakes consist of clumps of many individual crystals, tangled and frozen together. At very low temperatures, such clusters do not form and individual crystals fall as powder snow, which is highly regarded by skiers and also by transport authorities, because it is easily shifted by snow blowers. Snow that forms at temperatures just below freezing ('wet snow') causes major transport problems because it often sets into solid ice under the slightest pressure and has to be removed by mechanical or chemical means.

A fresh snowfall contains a lot of air, but the snow gradually becomes compacted and larger crystals grow at the expense of smaller ones. This may lead to a layer of large crystals (known as depth hoar) that forms a weak layer at some distance below the surface. These conditions may lead to avalanches on steep slopes, where the uppermost layers slide over the weaker lower layers. Avalanches may be triggered by vibrations from natural or human activity or with the onset of a thaw.

Snow that has persisted throughout the summer is known as firn and it is usually covered by fresh snowfall during the following winter. Persistent snow gradually becomes compacted into glacier ice and eventually, when it reaches a low enough altitude, melts to become the source of mountain streams and rivers.

Mist and fog

Both mist and fog consist of water droplets suspended in the air. Technically, the term mist is used when visibility exceeds 1 km (3,280 ft) and fog, when visibility is less. Only when visibility is less than 200 m (650 ft) is ground transport seriously affected, so frequently when meteorologists prepare weather forecasts for the general public they use the term 'fog' only when visibility is reduced to this still lower value.

As might be expected, fog forms when moist air is cooled to the dewpoint. This is just the same process as found in clouds and mist and fog may be regarded simply as cloud at ground level. There are, however, specific terms for particular types of fog: upslope fog, advection fog and radiation fog. Hill, or upslope, fog forms when air is forced to move uphill until it becomes saturated. Advection fog forms when warm air moves over surfaces that are colder than the air itself – surfaces such as the cold sea or snow- or ice-covered ground – and is consequently cooled below the dewpoint. The resulting fog may then be carried over nearby areas. This is often the case with sea fog

▲ Rocks covered by a heavy snowfall. Snow on upland areas is a primary source of water for many of the world's major rivers.

generated over water that has been brought to the surface by upwelling of cold bottom water, as occurs off California, Namibia and Chile, among other places. The Namib and Atacama deserts are so dry that sea fog that drifts inland is a major source of water. The Grand Banks off Newfoundland are notorious for the fog created when warm maritime tropical air encounters the cold Labrador Current.

The most common form of fog is radiation fog, which generally forms on still, clear nights. The ground radiates heat to space and cools the air immediately above it. The lowermost layer is colder than air above it and thus tends to remain in place, becoming still colder until it reaches dewpoint and mist begins to form and later may thicken into true fog. Such fog commonly forms along streams and rivers and in the bottom of valleys. The absence of any wind is important, because even a slight breeze will cause turbulence, mixing the air to a substantial depth and thus delaying cooling.

Fog is densest and most extensive around dawn after the long period during the night when the Earth's surface cools. If during the morning some heat from the Sun manages to warm the ground, it will start to raise the temperature of the fog and gradually the fog droplets will evaporate and the fog start to thin. Some signs that fog will clear are a significant brightening in the early morning, a rise in the wind speed and patches of sky becoming visible directly over-

▶ Advection fog: sea fog generated over the California Current drifting onshore at Big Sur, central California.

head. The fog will lift slowly into low sheets of cloud (i.e., stratus) and if this occurs before midday it is likely that the fog will break up and disperse in the afternoon. The thicker the fog, the longer it lasts, and on some days in winter heating may be insufficient to disperse it, so it lasts all day, especially in valleys and above rivers. Signs that fog is likely to last are no brightening during the morning, a lack of wind and a generally heavy (i.e., humid) feel to the air.

In coastal regions in late winter, spring and even into early summer, fog will often persist all day. Sea fog is often very dense and pushes far inland when the prevailing wind is onshore. It lasts as long as the warm, humid and stable air stream persists and is cooled to the dewpoint by contact with the cold sea. But, as mentioned earlier, strong winds will break up fog by causing turbulent mixing throughout a deep layer of air. In coastal areas, the land breeze at night often carries sea fog back out to sea. Because the humidity of the air may vary considerably from place to place, fog is often patchy, especially early in the night. Such conditions may be particularly dangerous for drivers, because visibility may suddenly change over a short stretch of road.

Steam fog

When cold air flows above warm water, water vapour rises into the lowest layer of air, where it condenses to give a steam fog. Such tendrils of 'steam' are often seen rising from wet roads heated by the Sun. The same process often occurs on a large scale in arctic regions, where the phenomenon is known as arctic sea smoke. Generally, however, the air in polar regions is so cold and dry that the lowermost level is very unstable and the height of sea smoke is usually less than 10 m (33 ft).

On occasions, the water droplets in fog may be super-cooled (p.23) below the nominal freezing point of water. They then freeze immediately when they come into contact with any surface, forming a deposit of rime, which will be described shortly (p.104).

Dew

Dew, like radiation fog, generally forms under clear skies, when objects rapidly lose their warmth to space, cooling to the dewpoint. The layer of air beneath any leaves tends to act as an insulator, so they are normally colder than the surface of the ground itself, which acts as the source of

▼ *Tendrils of steam fog rising into a layer of cold air above a warm lake.*

▲ *Heavy early-morning dew on a golf course.*

moisture. In addition, some moisture may be deposited from the air itself, particularly on objects such as the roofs of cars, which rapidly become very cold.

The larger drops of water often seen at the very tips of leaves – especially grass – are not dew, but what are called guttation drops. Generally, plants transport large quantities of water through their leaves from which it evaporates into the air. When the humidity is high, the water cannot evaporate and accumulates at the very tips of the leaves as large drops.

Two optical phenomena are associated with dew. One is the dewbow, a rainbow-like effect that appears on areas of grass, particularly in autumn. Like a rainbow, it is created by light reflected by innumerable droplets of dew, suspended from spiders' webs, strung horizontally between the blades of grass. The second effect is the heiligenschein ('holy light'), which appears as a halo of bright light around the shadow of an observer's head. The dewdrops act to return light towards the source, in a way similar to the way in which 'cat's-eye' reflectors in the middle of the road function.

Frost

The term 'frost' is used both to indicate when temperatures drop below freezing and (most commonly) for the deposit of ice that forms on plants, the ground and other objects. In forecasts, two specific conditions are recognized:

air frost: temperature of 0°C (32°F) at a height of 1.25 m (4.1 ft)

ground frost: temperature of 0°C (32°F) at ground level

Frost may occur either when temperatures drop below freezing point through radiation to space on clear nights or when a cold air mass arrives. The formation of mist or fog on clear nights may delay (or even prevent) frost from occurring. If the air is very dry, ice may not be deposited on the ground until the temperature has fallen well below freezing, when the little moisture in the air finally condenses. If the water droplets in the air are supercooled (p.23), they freeze immediately they come into contact with a solid surface to give an opaque, white deposit known as rime. This freezing fog is particularly hazardous to drivers (among others), because – on turning a corner, perhaps – the windscreen may be instantly covered in an opaque sheet of ice. If there is a slight wind, rime builds up on the windward side of any object and may produce long 'feathers' that point into the wind.

Hoar frost is the deposit of rough, white ice crystals on objects that have cooled below freezing (normally by radiation of their heat to space). Some of the ice consists of dew

▼ *The rime-covered trees in the background clearly indicate that freezing fog was responsible for the ice, and that this is not heavy hoar frost.*

that has subsequently frozen and some of water vapour that has frozen from the air. In general, small objects cool faster than larger ones, so frost tends to form first on the edges of objects such as leaves and small twigs.

A slightly different form of ice coating is produced when rain falls on to a very cold surface. Such a situation often occurs in winter when a depression brings a warm front (with extensive nimbostratus clouds) that advances over a layer of polar air that has cooled the surface well below freezing point. Raindrops that hit the ground or other objects swiftly spread into a thin film of water before they freeze, producing a clear sheet of ice, known as glaze or glazed frost. A heavy layer may build up extremely rapidly causing great damage to trees, overhead wires and other structures. Such conditions created the destructive ice storm over parts of Canada and the north-eastern United States in January 1998.

When freezing rain falls on to road surfaces, the sheet of ice is normally almost completely transparent and appears dark, so it is commonly known as black ice. A road may seem merely to be wet, but is instead extremely treacherous: driving is very dangerous and even walking may be nearly impossible.

Soils, vegetation and microclimates

The nature of the soil and its vegetation at any locality have an influence on the weather. The ratio between the amount of radiation that falls on a surface and that which is reflected is known as the albedo, which is normally expressed as a percentage. This percentage varies greatly, depending on the exact nature of the surface, which governs how it heats up during the day. Some example albedoes are shown below.

Note the great difference between the albedo of water when the radiation just grazes the surface (70%) and when it is perpendicular to it (5%). As a comparison, the albedo of clouds is generally 50–60%. In general, of course, surfaces that lie more or less perpendicular to the incoming radiation will heat up more readily than those where the radiation

Albedo

Surface	Percentage
water (normal incidence)	5
forest	5–10
wet earth	10
rock	10–15
dry earth	10–25
grass	25
sand	20–30
old snow	55
water (grazing incidence)	70
fresh snow	80

makes a shallow angle with the surface. (This is, after all, the principal reason for the great difference in solar heating between the equator and the poles.)

Naturally, surfaces that are generally dark in colour will also tend to absorb more heat. Dark soil, for example, is naturally warmer than light-coloured soils and will therefore be the last to be affected by night frosts.

Air is an extremely efficient insulator and this means that any air trapped within the soil reduces the effect of daytime heating on the lower layers. Thus dry sand (for example) remains very cold deep down even though the surface may become extremely hot during the day. It also loses its heat very rapidly at night, because the air between the sand particles does not retain its heat as efficiently as water. This is why the surface in desert areas may be searingly hot during the day, but becomes extremely cold at night.

Water trapped in the soil reduces the effects of day-time heating and night-time cooling. During the day, for example, wet soils show little change in temperature. Both wet soil and vegetation evaporate water vapour into the air, producing a considerable cooling effect on the surface. As a result, dew and frost form first on lawns of gardens and shallow mists appear first over grass-covered areas, such as the flood-plains of streams and rivers. During the night, patches of mist or fog may be found alongside a road where the vegetation is thick and the soil is moist, even if the ground and the road itself are level and do not cross any water courses. In the clear areas between the patches of fog, the soil will be relatively bare and relatively dry.

As mentioned earlier, the exact slope and orientation of the surface – its exposure – together with the type of material forming the soil and the vegetation all act together to create an extremely wide range of microclimates, which affect which plants can be grown in any particular area. The protection offered by walls and wind-breaks, for example, may make a great difference to the types of plants that will flourish, while slopes allow cold air to drain away to a lower location, making it less likely that plants on them will be damaged by frost. Sometimes human intervention may completely alter a natural situation. The Rickmansworth Frost Hollow – where severe frosts are far more common than in the surrounding areas – in Hertfordshire, southern England, did not exist until a railway embankment was built across the valley, preventing the cold air from draining away. As a result a pool of extremely cold air regularly forms in the valley and some of the coldest temperatures for southern England are now recorded here.

One effect of vegetation that is of great importance to many individuals is its production of pollen. Although sufferers from hay fever may be allergic to pollen from many plants, that from grasses is frequently the most irritating.

The amount of grass pollen released into the air is greatest during early summer when the Sun warms grasslands on days when the air is unstable. Convection in the form of thermals carries the pollen up to the stronger winds that exist above the boundary layer, where it is carried over the surrounding areas, remaining suspended until convection dies down, which is typically around five or six o'clock in the afternoon. This is the peak time for pollen, when the daily pollen count reaches its maximum. The highest pollen counts are normally encountered when moderate winds bring dry weather. The lowest layers of air are unstable, because of the high surface temperatures, so convection readily disperses the pollen.

Certain trees also release large amounts of pollen into the air. In Scandinavia and Canada, for example, both birch and pine pollen may be present in great amounts. It is not uncommon for the concentration of pollen to be so great that it creates coronae (p.58) around the Sun and Moon. Most meteorological services now issue warnings when the pollen count is expected to be particularly high.

Rain washes pollen from the air, so frontal systems and showers usually bring some relief to sufferers. Coastal areas tend to be free from pollen – or have lower pollen counts –

▼ Bright yellow pollen from pine trees blanketed the ground and vehicles in North Carolina during a period of above-average temperatures which coincided with a lack of rain.

because of the onshore, pollen-free sea breezes. Depending on the amount of convection in the atmosphere, and whether there are any inversions, higher altitudes also tend to have clearer air.

Towns and cities: pollution

The most noticeable effects of human activity on the weather are to be found in built-up areas. Large towns and cities are usually warmer than the surrounding countryside. This heat island effect interferes with the free exchange of heat and moisture and greatly affects the wind. During the day, the concrete jungle, with its high thermal capacity, absorbs heat from the Sun, which it re-radiates slowly later, acting rather like a giant storage heater. Cities are thus significantly warmer than surrounding districts, often by as much as 5 deg C (9 deg F). This is why both fogs and frosts are less frequent in cities.

By day, a city is often a source of strong thermals, created by the excess energy released into the atmosphere. The higher temperatures and rising thermal currents create an artificial heat low – a low-pressure area caused by the rising warm air. The air surrounding the city is therefore at a slightly higher pressure and starts to converge on the city to replace the rising air. This establishes a weak circulation, and when the air is unstable and moist, localized town showers may develop. When, by contrast, there is a general inversion that acts as a 'lid' to convection, creating a layer of cloud such as stratus or stratocumulus, the increased turbulence over the city may be just enough to break up the cloud and produce a warm, sunny microclimate over the built-up area.

Winds in a city exhibit strange behaviour. Obviously, because of the presence of so many walls and buildings there are places that are sheltered from the wind. Courtyards and gardens may enjoy calm conditions, whereas out in the suburbs and open countryside the wind is quite strong. On the other hand, the lines of high buildings tend to funnel the wind along the streets and may cause it to accelerate through smaller and smaller gaps. At a crossroads in a city the wind may gust to gale force, create swirling clouds of litter, while the gradient wind outside the city may be no more than a gentle breeze. Similarly, the obstacles presented by large buildings produce major eddies and consequently are responsible for violent gusts of wind from seemingly arbitrary directions.

In high summer, cities may be subject to oppressively hot conditions and serious pollution. When pressure is high overhead and the upper air is warm and sinking, the lowest layers of air do not mix with the higher layers of air and are not diluted by them. Smoke from chimney stacks emerges almost horizontally, clearly demonstrating the stability of the surface air. Gentle eddies in the air gradually pull such pollu-tion down to street level, where it combines with another major source of pollution: the exhaust from vehicles. The concentrated burning of fossil fuels injects more and more carbon dioxide, sulphur dioxide, dust and other pollutants into the surface layers of air and air quality falls rapidly. The heat from the ground is trapped by the pollution haze and the temperature rises even farther. The sky often turns hazy, sometimes brownish in tint, but occasionally a dazzling milky-white and the visibility drops dramatically. This type of weather usually breaks down only when the temperatures rise so high that convection breaks through the 'lid' and the resulting heavy showers or thunderstorms wash away the

▼ *A vast swathe of eastern China covered by haze in June 2007, as imaged by NASA's Terra satellite.*

pollution. A change in the pressure pattern, with rising winds, will also disperse the haze, pushing the plume of pollution downwind of the city.

Another type of pollution may occur in damp, foggy weather. Under such conditions, smog (smoke fog) may be created from the waste products of burning fossil fuels, often arising from both industrial and domestic sources. The waste products find their way into the fog and become chemically active, in particular the sulphur dioxide reacts with water to form sulphuric-acid droplets. This pollutant – otherwise known as 'acid rain' – slowly attacks buildings, vegetation and the respiratory system in human beings. Even though recent years have seen measures taken in many countries to reduce sulphur dioxide emissions, levels in urban areas are still higher than in the unpolluted, natural environment, where decaying vegetation is the source.

Dry hazes have become a major problem in certain countries, such as India and China, where, in addition to industrial sources, there are many small domestic sources of smoke and other pollutants. Satellite images often show vast swathes of countryside hidden below dense hazes. Similar smoke hazes arise from wildfires (whether set by lightning or human agency) and agricultural burning (especially the slash-and-burn methods used in many countries). This pollution has also been shown to reduce rainfall at moderate altitudes, bringing more problems to areas where water is already in short supply. Although the haze does have the effect of reducing heating by sunlight over the areas affected – and thus slightly offsets global warming – it is a major health hazard that needs to be eliminated.

It is quite common for the column of smoke from forest or other fires to be capped by a cumulus cloud and these are sometimes known by the somewhat ugly term of 'pyrocumulus'. Occasionally, such clouds break away from the smoke column and are carried downwind. They can usually be recognized by the brownish, smoky tint to the bottom of the clouds, which contrasts sharply with the brilliantly white cloud droplets at the top of the cloud. With extremely fierce wildfires, the convection may be great enough to produce showers of rain or even a major thunderstorm. Lightning from such storms has been know to initiate further forest fires some distance away from the initial fire.

Some widespread hazes consist of photochemical smog. This primarily arises in urban areas when vehicle exhaust emissions, mainly hydrocarbons and nitrogen oxides, are exposed to sunlight. Various photochemical reactions take place, creating a number of noxious substances, including ozone, sulphuric acid and sulphates, all of which are deleterious to health. Although these substances are short-lived, dispersing after nightfall, they remain a significant health hazard. Although measures have been taken to reduce vehi-

cle emissions, in particular, a number of sources remain and, perhaps surprisingly, barbecues make a very significant contribution to such pollution. The situation is particularly serious when a low-level inversion, often in combination with hills surrounding the urban area, traps the pollutants and prevents them from being dispersed over a wider region. This was the case in the Los Angeles area, where smog first became recognized as a major problem.

A slightly different type of result of human activity is found in the plumes from the cooling towers of power stations. These are often mistaken for plumes of smoke, but they simply consist of water vapour that has condensed in the cooler atmosphere. They have, however, been known to contribute to the formation of cumulus clouds above power stations and have even, when conditions have been significantly unstable, contributed to the formation of cumulus congestus clouds that have produced a brief shower of rain.

▼ *A power station in West Virginia producing plumes of steam from its cooling towers and columns of smoke from its chimneys, high above radiation fog that formed overnight.*

NORTH AMERICAN WEATHER

Much of the weather of North America is affected by the major barrier to the westerly winds that is formed by the Rocky Mountains and their subsidiary ranges. These mountains capture much of the moisture from the maritime tropical air from the Pacific, creating much drier conditions over the Great Plains. Westerlies crossing the mountains may give rise to the famous Chinook (the 'snow-eater'), the föhn wind on the leeward side, which has been known to raise temperatures by as much as 27 deg C (49 deg F) in just five minutes. Depressions often weaken over the mountain ranges, but reform to the east, becoming very strong, move south-east and then north-east, greatly influencing the weather over the Great Plains and the eastern states. Off the Pacific west coast, warm air streaming over cold upwelling water often (particularly in summer) produces extensive sea fog that then drifts inland. Similarly, on the east coast, warm moist air passing over the cold Labrador Current produces heavy fog over the Grand Banks off Newfoundland, but otherwise the Atlantic seaboard has more of a continental than a maritime climate.

Arctic air that builds up over the Canadian Arctic islands and continental polar air (from northern Canada) often sweep down through the central and eastern regions, bringing cold air as far as the Gulf states. This air stream is often strengthened by the cold highs that develop over the prairies in winter. Arctic-Front depressions frequently move south-east, merging with Polar-Front depressions, bringing severe blizzards to the Great Plains, the Great Lakes and the eastern seaboard. In early winter, the Great Lakes them-

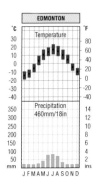

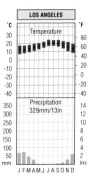

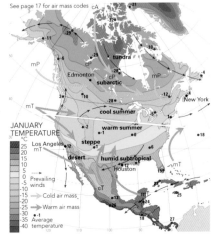

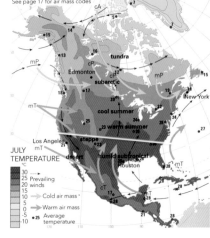

selves add significant amounts of moisture to the air, leading to especially heavy snowfalls to the south and east.

In the south in summer, maritime tropical air penetrates the valley of the Mississippi and even into Canada. This unstable air is the main cause of the violent thunderstorms and tornadoes experienced in the central states. This is accentuated when dry, continental tropical air flows from the hot south-western states and over the moist air from the Gulf of Mexico.

Depressions generated at the Polar Front over the Pacific often decay before they reach the land. This contrasts with the situation in the east, where the Polar Front may occasionally reach as far south as the Gulf of Mexico. Depressions generated at low latitudes grow as they travel up the East Coast, feeding on the temperature difference between cold continental air over the land and the warm maritime air over the Atlantic. They are steered on their course by the Appalachians, while depressions formed over Colorado follow a parallel track farther west. Destructive hurricanes generated over the warm southern North Atlantic sometimes follow a similar curved track up the eastern seaboard.

The January and July charts include indications of the location of the various natural climate zones. Brief details of these are:

tundra: treeless polar regions, very cold winters

subarctic: forested (taiga), moderate precipitation, cold winters

cool summer: moderate precipitation, cold winters, but no high summer temperatures

warm summer: moderate precipitation, cold winters

steppe: semi-arid with little precipitation, cold winters, warm or hot summers

desert: very low rainfall, extremely hot summers

humid subtropical: hot or warm throughout the year, high humidity year-round, heavy rainfall

Mediterranean: wet cool winters, hot dry summers

west coast: moderate rainfall year-round, no great temperature extremes, changeable weather

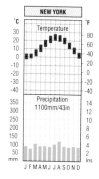

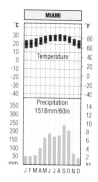

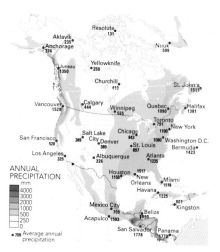

EUROPEAN AND MEDITERRANEAN WEATHER

The weather over Europe, especially western Europe, is notoriously variable, because it is affected by nearly every type of air mass, although the east has a more truly continental climate. Frequent low-pressure systems travel in from the warm Atlantic and may penetrate far into the continent, especially in winter. Blocking anticyclones are fairly frequent, particularly over Scandinavia, and these deflect the depressions northwards over Iceland or south-east and north of the Alps. The occurrence of blocking highs is very variable, so they sometimes bring continental polar (or even Arctic) air and exceptionally severe weather to western Europe. At other times they bring either fine, warm or hot weather, or seemingly endless dull days when depressions linger over the area. High-pressure patterns may also bring waves of cold, but fairly dry, maritime polar air.

Western Europe has much warmer winters than might otherwise be expected for its latitude, thanks to the warm waters of the North Atlantic Drift (often, but erroneously, called the Gulf Stream). This enables subtropical and tropical plants to be grown in sheltered parts of south-west Scotland, for example, at a latitude that corresponds to that of Labrador on the opposite side of the Atlantic. Despite this, warm moist maritime tropical air from the Atlantic encounters cold sea surfaces off the western coasts producing extensive sea fog. Maritime tropical air usually forms the warm sector of most depressions. When it encounters cold polar or continental arctic air from the east, considerable snowfall may occur.

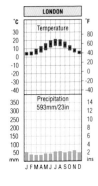

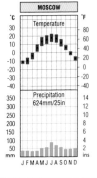

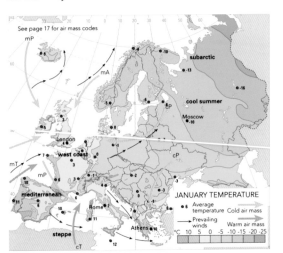

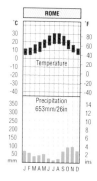

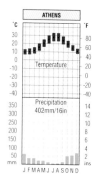

The Scandinavian mountains produce significant differences in the climate on the two sides of the range and both they, and the Alps, exert considerable steering effects on the paths of depressions. Low pressure to the north of the Alps may draw warm Mediterranean air over the mountains and give föhn conditions on the northern side.

The Mediterranean itself has a distinct climatic regime, to which its name has been given. This is characterized by considerable winter rainfall, followed by hot, dry summers. It is dominated in summer by the subtropical high-pressure region and continental tropical air from Africa. In winter, the sea's warmth, a generally westerly air stream and an additional front (the Mediterranean Front) produce a series of depressions that travel eastwards along its length. Such depressions travelling east are often regenerated at the eastern end of the Mediterranean by continental polar air from western Russia and then tend to swing north-east over the Black Sea. The complicated relief on the northern side of the Mediterranean gives rise to considerable modifications by local effects. The mountains, too, are responsible for the strong katabatic winds (p.94) of the Mistral and the Bora (the latter affects the head of the Adriatic). From the African side, the Sirocco and the Khamsin winds bring hot, continental tropical air northwards, where it may pick up considerable moisture from the sea.

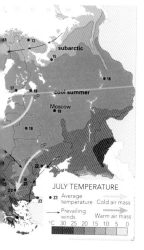

115

WEATHER OVER AUSTRALIA AND NEW ZEALAND

Australia is sufficiently far north for it to come well within the influence of equatorial air and indeed the north-western coast (in particular) is subject to a monsoon regime with north-westerly winds in summer and south-easterlies in winter. Hot and humid equatorial and maritime tropical air readily penetrates over the central desert regions as far as 21°S, and may bring rains even farther south. However, the mountains of the Great Dividing Range in the east run along the complete length of the coast and prevent easterly maritime tropical air from penetrating very far inland. The effects of the Range are seen in the maritime climate of the coastal strip while to its west the climate is much drier and of a more markedly continental type.

The interior itself is a source region for continental tropical air and the air mass that develops is warm and dry in winter. In summer, the intense heat produces a very hot, dry air mass which – in the form of a very hot, dry and dusty wind – is known as the Brickfielder.

In the south, maritime polar air is the predominant influence, with a constant series of depressions in winter bringing rain to the southern coasts. In summer, the general southwards migration of the dominant pressure areas causes these depressions to follow tracks at higher latitudes, so only Tasmania is occasionally affected. These southern regions, then, have a Mediterranean-type climate, with significant

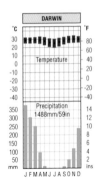

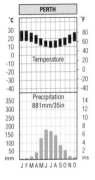

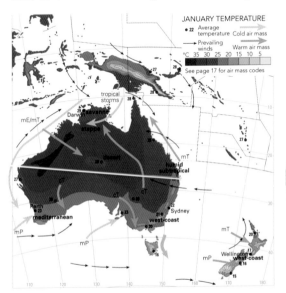

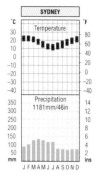

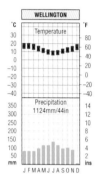

amounts of rain in winter, but dry, warm summers. Sudden changes of wind direction from northerly to southerly with incursions of very cold air affect the south and south-east.

Destructive tropical cyclones may affect the north of the continent, especially in areas close to the Arafura and Timor Seas. It is not unknown for these storms to follow long tracks right round the western side of the continent. The North Island of New Zealand, too, may feel the effects of similar storms that originate in the western Pacific.

Because it is far from any major land mass, New Zealand does not experience the influence of continental air and the dominant types are maritime tropical and maritime polar air. These are present in the endless series of low-pressure systems that cross the country with the prevailing westerly winds, the Roaring Forties, encircling the globe between 40°S and 50°S. The relief, particularly of South Island, leads to heavier rainfall on the west coast and greater sunshine in the east, sheltered from the north-westerly winds. In the absence of continental influences there are only rare periods of settled weather, but the proximity of the Antarctic does mean that the maritime polar air is colder than comparable air in the northern hemisphere.

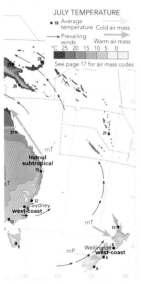

117

OBSERVING THE WEATHER

The preceding sections should give you a basis for predicting the likely course of the weather – at least on a local scale. A knowledge of cloud types and how to recognize stable or unstable conditions helps to forecast what is likely to happen over the period of a day and more specific understanding of how cumulonimbus clouds (for example) develop should enable you to tell whether you are likely to be drenched by a shower. Local conditions also have a great influence and once you have started to recognize, say, the onset of a sea breeze by day or conditions that produce a mountain wind by night, changes that previously seemed mysterious will be so no longer. The main point is to be aware of what is happening: rather than thinking just 'clouds' to progress to thinking 'summertime cumulus congestus, so we may get some rain later' or rather than just saying 'where did that fog come from?' to know that it is radiation fog that has built up overnight and is likely to persist unless the wind rises or daytime heating by the Sun becomes powerful enough to burn it off.

You will begin to see how your own observations fit into the larger picture of the weather situation, and how it may develop, if you compare your notes with professional forecasts and observational reports, many of which are now available over the Internet. The most useful information is provided by charts plotting actual observations, synoptic charts and weather-radar images.

Official weather forecasts are based on simultaneous surface observations from all over the world as well as vast amounts of data from many other sources, such as satellites and ocean buoys. As mentioned earlier, data from all over the Earth is required to prepare forecasts for just three days ahead. Increasing numbers of surface observations are obtained from automated stations, but reports from trained observers (many of whom are amateurs) are still obtained at many thousands of traditional stations. The World Meteorological Organization co-ordinates observations, data-processing and the transmission of data worldwide under a system known as the World Weather Watch. There are three main components: the Global Observing System, the Global Data-Processing Forecasting System and the Global Telecommunication System. The overall system is an outstanding example of international co-operation – which it must be, of course, given that the weather knows no boundaries.

The World Meteorological Organization (based in Geneva in Switzerland) lays down a standard set of routine observations carried out at surface stations across the world. The frequency of these observations varies according to the particular station. Observations are made hourly (on the hour) at most major airports and military airfields, for instance. Other stations may obtain observations at less frequent intervals: every three, six or twelve hours. All stations, regardless of their position on the globe, make observations at 00:00 and 12:00 UTC (Coordinated Universal Time) and the data obtained at these times are the primary input data for many of the computer forecasting programs.

▼ *A calm autumn morning after a sharp frost developed under clear skies overnight. What does the day have in store?*

Time

As with many other scientific fields, meteorological observations are obtained and reported in what is known as Coordinated Universal Time (UTC), which is calculated by the intercomparison of several atomic clocks maintained by various national standards organizations. UTC corresponds to the time at the Greenwich Meridian and is not altered by the application of Daylight Saving or Summer Time. The designation for the Greenwich time zone is 'Zulu' (abbreviated 'Z') and meteorological reports are often quoted as being obtained at 12:00 Z, for example.

Surface observations

The surface observations that are normally reported hourly, and their standard designations when describing plotting (p.128) are:

- total cloud amount – N
- wind direction
- wind speed
- air (dry bulb) temperature – TT
- dewpoint (wet bulb) temperature – TdTd
- horizontal visibility – VV
- pressure (corrected to sea level) – PPP
- three-hour pressure tendency – ppa
- cloud type – C_L, C_M and C_H
- cloud height
- present weather – ww
- past weather – W_1W_2

In addition, many stations will record and report other observations, such as precipitation (normally 12- or 24-hour totals) or duration of sunshine, on a daily, weekly or monthly basis. Monthly means of observations are reported by climatological stations, for instance. Other stations may be established on a temporary basis for specific investigations. Observations are also returned on a regular basis from merchant ships and commercial aircraft.

The development of new instrumentation and electronics means that many of these observations may now be made and reported automatically. Some (such as cloud type) still rely upon a human observer. Automatic weather stations are being deployed in increasing numbers and these are indi-

▼ *An automated weather station in Glacier National Park in the USA. An interesting cap cloud of orographic stratus covers the peak in the background.*

► *Examples of typical station plots for manned (left) and automatic (right) sites. The meaning of the figures and symbols will be explained later.*

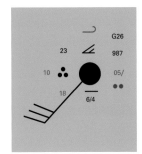

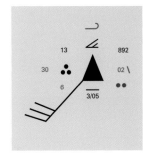

cated on station plots by different symbols. Automatic stations include the roadside installations that have now become a familiar sight, stations at remote and relatively inaccessible sites and oceanic buoys. The last group consists of moored buoys – including the buoys anchored in the middle of the Pacific Ocean to monitor conditions that lead to El Niño events – free-floating surface buoys and buoys ingeniously designed to float at a specific depth, which return to the surface at intervals to report observations, before descending automatically to their pre-set depth.

▼ *An early design of automatic weather buoy for reporting of inshore conditions.*

Upper-air observations

Important observations are obtained through the release of balloons at specific times. The simplest form, known as a pilot balloon, does not carry any instruments, but is inflated to give an approximately constant rate of climb. Tracking of the balloon – usually by observations every minute – enables the strength and direction of upper-air winds to be determined up to an altitude of approximately 15 km (50,000 ft) or more. Smaller balloons are sometimes used to determine the height of the cloudbase.

Larger, instrumented balloons, known as radiosondes, may carry various instruments, most commonly recording pressure, temperature and humidity, with the observations being radioed back to the ground. These balloons – some of which are fitted with Global Positioning System (GPS)

receivers for the accurate determination of their location – are released from a worldwide network of stations four times a day, at 00:00, 06:00, 12:00 and 18:00 UTC. Wind direction and speed are generally recorded just twice a day (often with balloons with just basic telemetry, known as windsondes), whereas the full set of data is recorded at 00:00 and 12:00 UTC. These latter observations are of vital importance in the preparation of forecasts because (among other factors) they provide direct information on the environmental lapse rate (p.24) up to altitudes of approximately 20 km (about 65,000 ft). Other forms of radiosonde may carry different sensors, such as those to monitor ozone or radioactivity levels.

▲ An observer prepares to release a radiosonde from Eureka on Ellesmere Island in the Canadian Arctic. He is holding the instrument package in his right hand.

Satellite observations

Satellite observations are now an indispensable part of forecasting. There are two types of specifically meteorological satellite: geostationary and polar-orbiting satellites. Geostationary satellites orbit the Earth at an altitude of 35,900 km (22,300 mi), above the equator. At such a location, a satellite orbits the Earth once a day and thus remains over the same point on the surface. From the satellite's position, the curvature of the Earth means that it has a field of view of about 120°, rather than a full 180°. However, several satellites are stationed around the equator and provide overlapping coverage, the principal ones being Meteosat at 0°, GOES-E at longitude 75°W, GOES-W at 135°W and GMS at 140°E. These provide continuous surveillance of weather around the world, by both day and night. (All the satellites carry out observations in various channels at infrared wavelengths and only data obtained at visible wavelengths are unavailable at night.) The original Meteosat series, for example, provided full-disk coverage at 30-minute intervals in three channels: visible, infrared and a channel chosen to detect water vapour. Meteosat Second Generation series satellites monitor 12 channels at a higher resolution, and have a repeat rate of 15 minutes.

Proper coverage of high latitudes cannot, however, be obtained from geostationary satellites, so their observations are supplemented by data from polar-orbiting satellites. These are launched into much lower orbits – at altitudes of 800–1,000 km (500–620 mi) – that are highly inclined to the equator and thus carry them over high latitudes. Each satellite continuously observes the surface beneath it, which is therefore covered as a never-ending strip. Generally the orbits are Sun-synchronous, i.e., the orbital parameters are arranged so that the satellite returns to the same position relative to the surface at approximately the same time each day. The orbital period is typically about 90 minutes and as the Earth rotates below the orbit

▼ *A portion of an image from a Meteosat Second Generation satellite at 11:00 UTC on 4 April 2004. Thanks to the greatly improved resolution, details of the snow-cover over the Norwegian mountains and the leading edge of a depression advancing from the west are clearly visible.*

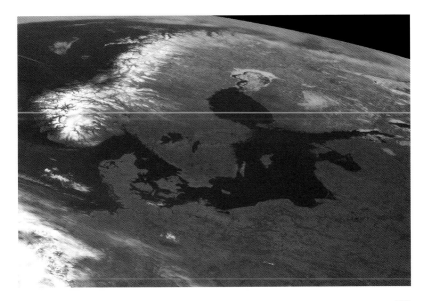

◀ A section of the continuous swathe observed by NOAA-14 on 7 May 2000, showing the British Isles relatively clear of cloud. This false-colour image was obtained by combining data from more than one channel to give approximately true colours.

the surface is covered by overlapping swathes, so that the whole Earth is monitored twice every day (once on a North-South pass, and once on a South-North pass twelve hours later). Again, the satellites observe at many spectral channels, providing data both day and night. The primary example of a series of such satellites is the Nimbus series, generally known as the NOAA satellites, after the United States' National Oceanic and Atmospheric Administration (NOAA), which provided them.

Apart from the great advantage of monitoring conditions over the oceans and sparsely inhabited regions of the Earth, where conventional surface observations (and observers) are sparse, satellite data may be used to derive a lot of useful information that is not available by any other means. Water vapour determinations have been mentioned, but other information includes wind directions, wave heights, cloud-top temperatures and many other factors that are required for the various types of forecast prepared nowadays.

Although not part of the meteorological observational network, many other satellites monitor the Earth, including the European Envisat environmental satellites and the NOAA Terra and Aqua satellites. Data from such satellites are not regularly used for forecasting, but are occasionally valuable in filling gaps that may occur in polar-orbiter coverage at the time of natural disasters, such as floods or tropical cyclones, as well as in individual studies of meteorological or climatological subjects.

Radar observations

Another form of observation that has become widely used in recent years and is likely to expand even farther in future is the use of dedicated weather radars. A number of countries employ such radars to provide continuous monitoring of precipitation and the data that they produce – together with the determination of lightning strokes, for example – has led to the development of 'nowcasting', the forecasting of short-term developments, i.e., changes likely to occur within the next six hours. Such forecasts are of particular value when severe weather is expected, including damaging hail, extreme thunderstorms, flash floods and tornadoes. The most sophisticated weather radars are Doppler radars, such as those deployed in the NEXRAD system across the United States, which are able to determine the motion of precipitation towards or away from the radar. Such information is invaluable in predicting the likely occurrence of tornadoes or the development of microbursts, extremely violent downdraughts of limited extent, which are exceptionally hazardous to aircraft that are taking off or landing.

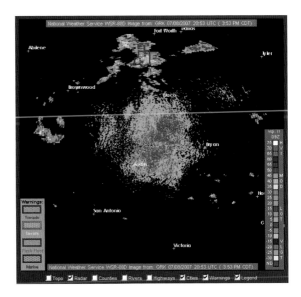

▶ A NEXRAD image from the Central Texas radar, showing general precipitation near the station, but intense precipitation towards Dallas and Fort Worth, to the north. The boxed area is under a flash-flood warning.

OBSERVING INSTRUMENTS, METHODS AND PLOTTING

Although increasing numbers of meteorological observations are made by electronic sensors, many – perhaps most – reporting stations still rely upon conventional instruments and methods, and a description of these is useful in understanding how observations are plotted and interpreted. The actual data are reported over the Global Telecommunication System in code, which is decoded to produce the plots.

Plotting observations

All station plots follow a standard format, where the position of the various figures and symbols are fixed and related to specific observations. The only exception is the indication of wind direction. The overall layout is shown in the diagram overleaf, together with samples of plots for both conventional, manned stations, where the station is shown as a circle, and automatic stations where the symbol is a triangle. The exact meaning of each of the separate figures or symbols plotted around the station symbol, and their colours, will be explained in the following sections and an example of a full chart is shown here. Conventionally, the colours used were black and red, but in computer-produced plots (where colours are used for land and sea) these are often shown as white and yellow, respectively.

It should be noted that certain data or symbols may be missing from individual station plots, either because cer-

Station plots

Station plots, many of which are now available over the Internet, give an extremely useful glimpse of the weather over a wide area at any one time, with the wind and present weather information being perhaps of most immediate interest. Unfortunately, because of various factors, large-area station plots tend to be rather variable in their content. The presentation depends greatly upon which national meteorological service has prepared the charts. Some do not differentiate between manned and automatic stations, while others use different symbols (such as stars) for automatic stations and others again use multiple colours for the different plotted parameters. The greatest limitations, however, are the fact that the stations included may vary from one plot to the next and that individual measurements may be missing from individual plots. This may be because the observations were not made or because there were errors in transmission. However, the symbols and codes used, and their positioning, are governed by international agreement, so at least those factors remain constant.

127

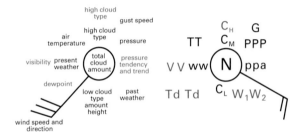

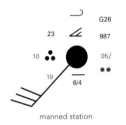 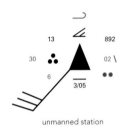

▲▼ *The standardized layout of data and symbols (top left) for observations, with the respective codes shown*

(top right). Also shown are a typical plot for a manned station (below left) and for an automatic station (right).

manned station

unmanned station

tain specific observations are not made at that station – cloud types at automatic stations, for example – or because the information is not available. The latter may occur because a specific observation is not possible – the high cloud type cannot be determined if the station is covered in low stratus or fog, for example – or because of instrumental or human error.

Total cloud amount (N)

In most countries, the total cloud amount is assessed in terms of the number of eighths (known as 'oktas') of the area of the sky covered by cloud. The principal exception is the United States, where the area is assessed in tenths. In practice, the differences are negligible, and the same set of eleven symbols is used, with slightly different numerical values in each system.

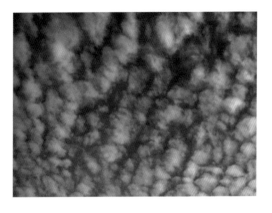

TOTAL CLOUD AMOUNT

Manned station	Automatic station	World (excluding USA)	USA
○	△	Clear (no cloud)	Clear (no cloud)
◐	◭	1 eighth	1 tenth or less
◔	◮	2 eighths	2–3 tenths
◑	◭	3 eighths	4 tenths
◐	◮	4 eighths	5 tenths
◓	◭	5 eighths	6 tenths
◕	◮	6 eighths	7–8 tenths
◑	◭	7 eighths	9 tenths or overcast with openings
●	▲	Overcast	Overcast
⊗	(No symbol)	Sky obscured by fog or other meteorological phenomena	Sky obscured by fog or other meteorological phenomena
⊖	◮	Sky obscured for other reasons or not observed	Sky obscured for other reasons or not observed

Note that for automatic stations, the last symbol is frequently encountered and indicates that no cloud data are available.

Wind speed and direction

Wind direction is measured with a simple wind vane, which nowadays normally provides an electrical readout, either on a direction dial or as a reading in degrees, measured clockwise from zero at true (not magnetic) North. The direction of a wind from the north itself is recorded as 360°. The value reported is the average value to the nearest 10°, taken over a few minutes. A flat calm is recorded as '000' and, for manned sites, plotted as a circle around the station circle. The wind direction (to the nearest 10°) is plotted by a 'shaft' running to the station symbol from the appropriate direction (with North at top). For an automatic station, a flat calm is shown by the omission of a wind shaft.

▲ *Manned station, seven-eighths cloud cover.*

Wind speed is measured by an instrument known as an anemometer. At official stations this is mounted at a standard height of 10 m (32.8 ft), which is regarded as the top of the boundary layer in which friction and turbulence are dominant. Where anemometers are mounted at a lower height (such as on oceanic weather buoys) a correction is applied to give the probable speed at the standard height. There are various forms of anemometer, one being the cup anemometer,

with three cups mounted on a vertical shaft, often above a wind vane. Another form, more common in the United States, is the propeller type, where a streamlined body carries a vane that turns the propeller at the front of the body into the wind. In most instances, the vertical or horizontal shaft is attached to a generator, which produces a current proportional to the wind speed. Other types of anemometer exist, including ultrasonic versions that are able to determine vertical velocity as well as the horizontal velocity normally recorded. (In violent updraughts and downdraughts, such as those found in cumulonimbus clouds, vertical motions may often exceed horizontal ones.)

▲ A common form of cup anemometer, mounted above a wind vane. The greater pressure on the concave side of the cups causes the vertical shaft to rotate with a speed proportional to the wind speed.

Wind speed is conventionally measured in knots (nautical miles per hour), but frequently converted to metres per second in meteorological investigations and reports:

1 knot = 0.514 m/s = 1.852 kph (1.688 ft/s = 1.15 mph)

On a plot, the speed of the wind is indicated by 'feathers' or 'barbs' added to the end of the shaft, speeds being rounded to the nearest 5 knots before plotting. Half-length feathers indicate 5 knots, full-length feathers 10 knots. A speed of 50 knots is shown by a triangle (occasionally called a 'pennant'). In the northern hemisphere the feathers are placed on the left, looking towards the station symbol. In the southern hemisphere they are placed on the right. Special symbols are used to indicate calm (already mentioned), variable direction and missing wind speed. Note that maximum gust speed is plotted separately. Gusts are also measured in knots and are plotted (to the nearest whole knot and preceded by the letter 'G') only when they exceed 25 knots.

▼ Wind direction 350°, 3–7 knots.

WIND SPEED AND DIRECTION

⭕ Calm	⦀ 23 – 27 knots	⦀ 53 – 57 knots	⦀ 83 – 87 knots
— 1 – 2 knots	⦀ 28 – 32 knots	⦀ 58 – 62 knots	⦀ 88 – 92 knots
⟍ 3 – 7 knots	⦀ 33 – 37 knots	⦀ 63 – 67 knots	⦀ 93 – 97 knots
⟍ 8 – 12 knots	⦀ 38 – 42 knots	⦀ 68 – 72 knots	⦀ 98 – 102 knots
⟍ 13 – 17 knots	⦀ 43 – 47 knots	⦀ 73 – 77 knots	✕ Wind direction variable
⦀ 18 – 22 knots	◢ 48 – 52 knots	⦀ 78 – 82 knots	✕— Wind direction given but wind speed missing

The Beaufort scale

Another method of specifying wind speeds uses the Beaufort scale of wind force. This was originally introduced by Rear-Admiral Francis Beaufort in 1806 and was adopted by the British Admiralty in 1838. Although it was originally described in terms of the sails that could be carried by a

▲ *An observational report from an automatic station, showing a gust speed of 27 knots.*

▶ *Rough seas with breaking waves and foam – a sea state corresponding to Near gale, Force 7 or Gale Force 8.*

BEAUFORT SCALE FOR USE AT SEA

Force	Description	Sea state	Knots	Kph
0	Calm	Like a mirror	below 1	below 2
1	Light air	Ripples; no foam	1–3	2–6
2	Light breeze	Small wavelets with smooth crests	4–6	7–11
3	Gentle breeze	Large wavelets; some crests break; a few white horses	7–10	12–19
4	Moderate breeze	Small waves; frequent white horses	11–16	20–30
5	Fresh breeze	Moderate, fairly long waves; many white horses; some spray	17–21	31–39
6	Strong breeze	Some large waves; extensive white foaming crests; some spray	22–27	40–50
7	Near gale	Sea heaping up; streaks of foam blowing in the wind	28–33	51–61
8	Gale	Fairly long and high waves; crests breaking into spindrift; foam in long prominent streaks	34–40	62–74
9	Strong gale	High waves; dense foam in wind; wave-crests topple and roll over; spray interferes with visibility	41–47	75–87
10	Storm	Very high waves with overhanging crests; dense blowing foam, sea appears white; heavy tumbling sea; poor visibility	48–55	88–102
11	Violent storm	Exceptionally high waves may hide small ships; sea covered in long, white patches of foam; waves blown into froth; visibility severely affected	56–63	103–117
12	Hurricane	Air filled with foam and spray; extremely bad visibility	≥64	≥118

		BEAUFORT SCALE FOR USE ON LAND		
Force	Description	Events on land	Knots	kph
0	Calm	Smoke rises vertically	below 1	below 2
1	Light air	Direction of wind shown by smoke, but not by wind-vane	1–3	2–6
2	Light breeze	Wind felt on face; leaves rustle; wind-vane turns to wind	4–6	7–11
3	Gentle breeze	Leaves and small twigs in motion; wind extends small flags	7–10	12–19
4	Moderate breeze	Wind raises dust and loose paper; small branches move	11–16	20–30
5	Fresh breeze	Small leafy trees start to sway; wavelets with crests on inland waters	17–21	31–39
6	Strong breeze	Large branches in motion; whistling in telephone wires; difficult to use umbrellas	22–27	40–50
7	Near gale	Whole trees in motion; difficult to walk against wind	28–33	51–61
8	Gale	Twigs break from trees; difficult to walk	34–40	62–74
9	Strong gale	Slight structural damage to buildings; chimney pots, tiles, and aerials removed	41–47	75–87
10	Storm	Trees uprooted; considerable damage to buildings	48–55	88–102
11	Violent storm	Widespread damage to all types of building	56–63	103–117
12	Hurricane	Widespread destruction; only specially constructed buildings survive	≥64	≥118

frigate, it was later adapted to be based on sea state and a version was devised for use on land. Speeds are specified in terms of Force, from Force 1 (light air) to Force 12 (hurricane). Shipping forecasts (in particular) frequently give expected wind speeds as 'Force 5' or 'Storm Force 10', for instance. In the tables, note that the scale is defined in terms of knots, so the values in kilometres per hour are approximate equivalents.

Temperature (TT)

Temperature is recorded with various standard forms of thermometer and a station will normally have four of these. One, known as the dry-bulb thermometer, is a mercury-in-glass thermometer which is mounted vertically and registers the current air temperature, which is normally read to the nearest 0.1°C (or 0.1°F). A similar thermometer (the wet-bulb thermometer) has a muslin covering over the bulb, with a wick that dips into a container of distilled water. Evaporation of water from the covering lowers the temperature registered by the thermometer. The difference between the dry-bulb and wet-bulb readings may be used to derive the humidity of the air (discussed in more detail shortly). These

▼ *Air temperature 16°C, dewpoint 13°C.*

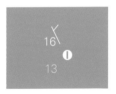

two thermometers are read whenever observations are made, whether hourly or at longer intervals, depending on the station. The current air temperature (TT), taken to the nearest whole degree Celsius, is plotted with two digits and a minus sign for temperatures below freezing, to the left of the station symbol.

Humidity (dewpoint – TdTd)

As just mentioned, humidity may be determined from the readings given by a pair of dry- and wet-bulb thermometers mounted in an instrument screen. When a fixed screen is not available, use is often made of a handheld device known as a whirling psychrometer, which consists of wet- and dry-bulb thermometers mounted together in a frame with a rotating handle, and which may be whirled round like a football rattle to give full ventilation. A device for registering humidity directly is known as a hygrometer, of which there are various forms. A recording type, called a hygrograph, is often used, where the sensing element is frequently human hair, which expands and contracts with changes in humidity. The changes are mechanically amplified and used to position a recording pen on a chart in a manner similar to that used in barographs and thermographs. As with current air temperature, the dewpoint temperature is plotted to the nearest degree Celsius (with two digits and a minus sign if necessary) in red or yellow.

Maximum and minimum temperatures

The other two thermometers are mounted almost horizontally, and are recording thermometers registering the maximum and minimum temperatures that have occurred since they were reset. Such thermometers are normally read and reset every 24 hours, the most common time for which is 09:00 local time. The standard form of maximum thermometer is mercury-in-glass, with a constriction in the neck, just above the bulb. The mercury column is free to expand beyond the restriction, but breaks when the temperature drops, thus enabling the maximum temperature to be determined. The thermometer is reset by shaking it so that the mercury is forced back through the constriction – just like old-fashioned clinical thermometers.

▼ *A typical small automatic weather station, recording wind direction and strength, temperature and precipitation.*

With calm conditions, the maximum temperature on any one day tends to occur in mid-afternoon. The timing may vary considerably, however, if a depression or other event occurs during the 24-hour period.

The minimum thermometer contains alcohol rather than mercury, because alcohol has a lower freezing point and so may be used at lower temperatures. Immersed within the alcohol is a light index, which does not move as the column expands with an increase in temperature, but which is dragged backwards when the alcohol column contracts. It thus records the minimum temperature attained since it was reset, which normally occurs at night, some time before dawn. Again, as with the maximum temperature, there may be significant variations if the prevailing air mass changes during the day. Resetting the index is normally carried out by tilting the thermometer so that the index slides along until it is in contact with the end of the alcohol column. A commonly seen form known as Six's thermometer, often sold for home use, consists of a U-shaped tube with two bulbs, one of which is filled with a clear liquid that expands and contracts with changes in temperature and acts on a short mercury column that carries an index at each end. The indexes are spring-loaded to remain in the maximum and minimum positions until reset, which is normally done by dragging them along with a magnet held on the outside of the tube.

In addition to the thermometers, some stations also make use of a thermograph, where changes in temperature cause variation in the curvature of a bi-metallic strip. This variation is amplified mechanically and transferred to a pen, which records the changes on a graduated chart, mounted on a slowly rotating drum. Automatic stations use thermocouples, where a change in temperature alters the resistance of an electrical circuit, thus giving a reading.

To ensure that the readings are accurate, the thermometers are mounted in a suitably designed screen that protects them from direct sunlight, but also ensures that there is a ready flow of air over them. At surface stations this often consists of an enclosure with a double roof and double-louvered sides. Such screens are often made large enough to accommodate additional instruments, such as a thermograph or hygrograph (for recording humidity). Automatic stations normally use a form of screen, originally known as a marine screen, which looks like an inverted stack of soup dishes.

Pressure (PPP)

Pressure used to be recorded by the classic form of barometer, consisting of a vertical tube sealed at one end, with the other held below the surface of a container of mercury. Once filled with mercury, the tube was temporarily closed, then inverted and the end uncovered beneath the level of

▼ *A modern version of the classic mercury-in-glass barometer, showing the reservoir at the base and the thermometer used for correcting the readings.*

the mercury in the reservoir. The mercury column would drop, leaving a vacuum at the closed upper end. The height of the column of mercury in the tube (approximately 75 cm (30 in)) is a direct measure of the weight of the atmosphere above the instrument and varies accordingly as atmospheric pressure changes. One disadvantage of this form of barometer is that any readings of the height of the mercury column (against graduations engraved on the glass tube) must be adjusted for the prevailing temperature. A thermometer is therefore frequently incorporated into the design.

Another form of barometer uses an evacuated metal capsule rather than a mercury column to sense pressure changes. These are known as aneroid ('without air') barometers, and are the type most commonly found in ordinary homes. Changes in air pressure cause the surfaces of the capsule to flex against an internal spring. In some instruments, the movement is amplified by using a stack of capsules mounted one on top the other. In home barometers, the motion is converted mechanically into movement of a pointer, giving the readings from a graduated scale. In precision aneroid barometers, such as those used by official weather stations, the reading is indicated by graduated drums with the correct position indicated by an electrical sensor.

Similar aneroid capsules are used in barographs, which continuously record atmospheric pressure over a period of a week on a strip of paper wrapped around a rotating drum. A barograph is particularly useful in that it immediately indicates the pressure tendency, i.e., whether pressure is

▲ *Current pressure 1,007.2 hPa.*

▼ *The form of barograph most commonly encountered, recording pressure on a narrow chart over a period of one week.*

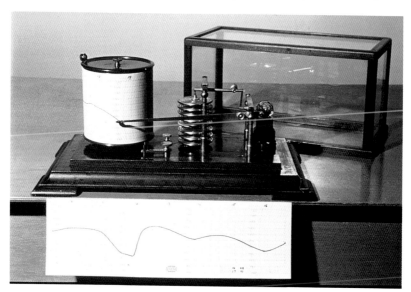

▲ *Pressure tendency over past three hours, 1.4 hPA overall change, rising then steady.*

increasing, decreasing or remaining steady, which is an important factor in determining how the weather is likely to change (and how rapidly) in the next few hours. Without a paper record, the pressure tendency must be obtained by comparing the latest measurement with ones obtained earlier. This is not a great problem when readings are made once an hour, but is impractical when readings are taken just once or twice a day.

Pressure is frequently measured in millibars (mb), or thousandths of a theoretical average sea-level pressure (called a bar). Meteorologists also specify pressures in terms of the hectopascal (hPa), because the basic unit (the Pascal, Pa) is rigorously defined scientifically, whereas the bar is not. In practice, however, one millibar (1 mb) is identical to one hectopascal (1 hPa). Radio and television broadcasters tend to give pressures in millibars, because these are more familiar to the general public. Although some pressure readings are still given in 'inches of mercury' – and such values are often found on the scale of home aneroid barometers – the official unit, even in the United States, is the millibar or hectopascal.

Pressure readings are normally taken to the nearest 0.1 hPa (0.1 mb). Before such readings may be used, however, a correction must be applied for the height of the barometer above sea level. Pressure decreases by approximately 1 hPa for every 10 m (32.8 ft) of altitude, so an appropriate amount must be added to any reading to obtain the theoretical sea-level value (unless the station is below sea level, when the correction must be added to the reading). Corrected sea-level pressures may then be plotted to create an isobaric chart. In passing, it may be noted that the average sea-level pressure is not actually 1,000 hPa (i.e., 1,000 mb or one bar), but instead 1,013.25 hPa.

Technically, pressure readings should also be adjusted to take account of the variation in the strength of gravity at different points on the Earth. For our purposes, however, this correction may be ignored. Pressure (PPP) is plotted in hPa (or millibars) to the nearest tenth of a millibar, and as three digits. A pressure of 991.7 is plotted as 917 and one of 1,007.3 as 073.

Pressure tendency (ppa)

The pressure tendency is specified as the amount and nature of the change in the pressure readings taken over the past three hours. The change over three hours is reported (in hPa), together with the form of change (steady, rising, falling, rising then falling, etc.). The change over that period is given in tenths of a hectopascal and plotted (in red or yellow) as two digits, so '30' would indicate a change of 3 hPa. Eight symbols are used for plotting the form of change, black/white for rising overall and red/yellow for falling overall.

PRESSURE TENDENCY

⟋	Rising, then falling
⟋	Rising, then steady
⟋	Rising
⟍	Falling, then rising
⟍	Falling, then rising
⟍	Falling then steady
⟍	Falling
⟋	Rising then falling

▲ *Low cloud:*
cumulonimbus;
medium cloud:
altocumulus from
spreading cumulus
or cumulonimbus;
high cloud: dense cirrus.

Cloud types (C$_L$, C$_M$ & C$_H$)

The recording of cloud types and the symbols used for plotting are based on the division into low (C$_L$), medium (C$_M$) and high (C$_H$) clouds described earlier (p.27). The low cloud symbol is plotted below the station symbol, the medium cloud immediately above and the high cloud above that (in red/yellow). With dense low cloud, the higher cloud observations may, of course, be impossible and the symbols will be missing.

LOW CLOUD TYPE (C$_L$)

Cumulus with little vertical extent

Cumulus of moderate or strong vertical extent

Cumulonimbus without fibrous or anvil top

Stratocumulus formed by the spreading out of cumulus

Stratocumulus not formed by the spreading out of cumulus

Stratus in a more or less continuous sheet or layer

Stratus fractus of bad weather

Cumulus and stratocumulus at different levels

Cumulonimbus, fibrous or anvil top

MEDIUM CLOUD TYPE (C$_M$)

Altostratus through which the sun or moon may be weakly visible

Altostratus, dense enough to hide the sun or moon, or nimbostratus

Altocumulus, the greater part of which is semi-transparent and at a single level

Patches of altocumulus, the greater part of which is semi-transparent occurring at one or more levels

Semi-transparent altocumulus in bands or altocumulus in one or more fairly continuous layers, progressively invading the sky

Altocumulus resulting from the spreading out of cumulus (or cumulonimbus)

Altocumulus in two or more layers, not progressively invading the sky or altocumulus together with altostratus or nimbostratus

Altocumulus with sproutings in the form of small towers or battlements (altocumulus castellanus)

Altocumulus of a chaotic sky, generally at several levels

Note that thick altostratus and nimbostratus are shown by a single symbol.

HIGH CLOUD TYPE (C_H)

⌐	Cirrus in the form of filaments, strands or hooks, not progressively invading the sky
⌐⌐	Dense cirrus patches, which do not increase and seem to be the remains of the upper part of cumulonimbus; or cirrus with sproutings in the form of small turrets or battlements (cirrus spissatus or cirrus castellanus)
⌐	Dense cirrus, often in the form of an anvil; the remains of the upper parts of cumulonimbus (cirrus spissatus)
⟋	Cirrus in the form of hooks or of filaments, or both, progressively invading the sky; they generally become denser as a whole
⌐	Cirrus and cirrostratus, or cirrostratus alone; progressively invading the sky, but not reaching 45° above the horizon
⟋	Cirrus and cirrostratus, or cirrostratus alone; progressively invading the sky, reaching more than 45° above the horizon, but without the sky being totally covered
⌐	Veil of cirrostratus covering the celestial dome
⌐	Cirrostratus not progressively invading the sky and not completely covering the celestial dome
⟋	Cirrostratus alone, or cirrocumulus accompanied by cirrus or cirrostratus or both, but cirrocumulus is predominant

Cloud height

Cloud height was formerly determined by estimation, which obviously required considerable experience to be even approximately accurate. At certain stations measurements could be made by tracking a pilot balloon and determining when it entered the cloudbase. Modern determinations are made with a cloudbase recorder (sometimes known as a ceilometer). Such instruments normally take the form of a lidar – a laser detection and ranging system – where timing of the laser pulses reflected from the cloudbase enables its altitude to be determined with considerable accuracy. The coding of cloud heights differs between automatic and manned stations, with height being given in hundreds or thousands of feet, respectively, as shown in the height tables for both manned and automatic stations, on the right.

On plots, the amount of low cloud in oktas (or tenths), expressed as a number, is shown beneath the low-cloud symbol. It is followed by a slash (/) and then the code for the cloud height.

▲ Low cloud amount is 4 oktas, and height is 1,000–1,999 ft.

AUTOMATIC STATIONS

Code	Height in ft
00	< 100
05	500
10	1,000
15	1,500
20	2,000
30	3,000
40	4,000
50	5,000
60	6,000

MANNED STATIONS

Code	Height in ft
0	0–149
1	150–299
2	300–599
3	600–999
4	1,000–1,999
5	2,000–2,999
6	3,000–4,999
7	5,000–6,499
8	6,500–7,999
9	8,000 or above
/	height unknown

Visibility (VV)

Visibility is conventionally assessed by determining whether objects at known distances from the station may be seen. (At night, lights are normally used as targets.) Automatic visibility meters measure the intensity of a light over a relatively short distance and this intensity is related to the conventional scale of distances.

Because low visibilities are more significant, in that they pose considerable danger – particularly for vehicles, ships and aircraft – a fine scale is used for distances up to 5 km (3.1 mi) and a coarser one for distances between 5 and more than 70 km (3.1 and over 43.5 mi). Two digits are plotted. Up to 5 km (3.1 mi) these represent the visible distance in tenths of a kilometre, with the decimal point omitted. The codes 51 to 55 are omitted. Between 5 and 30 km (3.1 and 18.6 mi), distances are given to the nearest kilometre, and the code value is obtained by adding 50. Thus, '56' represents 6 km (3.7 mi) and '80' represents 30 km (18.6 mi). Above 30 km (18.6 mi), the scale increases in 5-km (3.1-mi) increments, '81' representing 35 km (21.7 mi), '88' equalling 70 km (43.5 mi). Code '88' is used for distances about 70 km (43.5 mi). These codes are summarized in the abbreviated tables shown. The code is plotted (in red or yellow) on the far left of the station symbol.

▼ *Visibility is coded (yellow) as '82' representing 40 km (24.9 mi).*

CODES FOR DISTANCES BELOW 5 KM	
Code	**Distance (km)**
00	< 0.0
01	0.1
02	0.2
–	–
–	–
–	–
50	5.0

CODES FOR DISTANCES ABOVE 5 KM	
Code	**Distance (km)**
56	6
57	7
58	8
–	–
–	–
–	–
80	30
81	35
82	40
–	–
–	–
–	–
88	70
89	> 70

Present weather

The information given by the 'present weather' code or symbol covers a wide range of conditions. It is one of the most useful items shown by station plots although, regrettably, it is often missing from individual hourly plots. Nevertheless, it enables one to gain an immediate mental picture of the weather prevailing at that particular time, which gives you a good indication of how it is likely to develop. The coding seems complicated, because it covers no fewer than 100 different conditions, but may be broken down into various groups. Don't try to memorize all the different symbols.

Once you have examined a few station plots, you will find that certain common symbols are readily recognized, and you can look up the rarer ones in these lists.

CODES

WW Codes 00–19: No precipitation, fog (except for 11 and 12), duststorm, sandstorm, drifting or blowing snow at the station at the time of observation, nor (except for 09 and 17) during the preceding hour.

Symbol	Code	Description
	00	Cloud formation not observed or observable
	01	Clouds dissolving or decreasing
	02	State of sky generally unchanged
	03	Clouds generally forming or developing
⌒	04	Visibility reduced by smoke haze
	05	Haze
S	06	Widespread dust in the air, not raised by wind at or near the station at the time of observation
$/	07	Dust or sand raised by wind at or near the station at the time of observation, but not well developed dust whirl(s), and no sandstorm seen; or, in the case of ships, blowing spray at the time of observation
ẞ	08	Well developed dust whirl(s) or sand whirl(s) seen at or near the station during the preceding hour or at the time of observation, but no duststorm or sandstorm
(S)	09	Duststorm or sandstorm within sight at the time of observation, or at the station within the preceding hour
=	10	Mist
≡ ≡	11	Patches of shallow fog or ice fog
≡ ≡	12	More or less continuous shallow fog or ice fog less than 2 m on land or 10 m at sea
⦉	13	Lightning seen, no thunder heard
⦁	14	Precipitation within sight, not reaching the ground or surface of the sea
)•(	15	Precipitation within sight, reaching the ground or surface of the sea, but distant, i.e., estimated to be more than 5 km from the station
(•)	16	Precipitation within sight, reaching the ground or surface of the sea, near to, but not at the station
☈	17	Thunderstorm, but no precipitation at the time of observation

Symbol	Code	Description
▽	18	Squalls at or within sight
)(	19	Funnel cloud(s) at or within sight of the station during the preceding hour or at the time of observation

WW Codes 20–29: Precipitation, fog, ice fog or thunderstorm during the preceding hour, but not at the time of observation

Symbol	Code	Description
,]	20	Drizzle (not freezing) or snow grains, not falling as showers
•]	21	Rain (not freezing), not falling as showers
✳	22	Snow, not falling as showers
•✳]	23	Rain and snow or ice pellets, not falling as showers
~]	24	Freezing drizzle or freezing rain, not falling as showers
•▽]	25	Shower(s) of rain
✳▽]	26	Shower(s) of snow, or of rain and snow
△▽]	27	Shower(s) of hail, or of rain and hail
≡]	28	Fog or ice fog
℞]	29	Thunderstorm (with or without precipitation)

WW Codes 30–39: Duststorm, sandstorm, drifting or blowing snow

Symbol	Code	Description
⊖→|	30	Slight or moderate duststorm or sandstorm, has decreased during the preceding hour
⊖→	31	Slight or moderate duststorm or sandstorm. No appreciable change during the preceding hour
|⊖→	32	Slight or moderate duststorm or sandstorm, has begun or increased during the preceding hour
⊜→|	33	Severe duststorm or sandstorm, has decreased during the preceding hour
⊜→	34	Severe duststorm or sandstorm. No appreciable change during the preceding hour
|⊜→	35	Severe duststorm or sandstorm, has begun or increased during the preceding hour
↓→	36	Slight or moderate drifting snow, generally low (below eye level)
⇊→	37	Heavy drifting snow, generally low (below eye level)
↑→	38	Slight or moderate drifting snow, generally high (above eye level)
⇈→	39	Heavy drifting snow, generally high (above eye level)

Symbol	Code	Description
		WW Codes 40–49: Fog or ice at the time of observation
	40	Fog or ice fog at a distance at the time of observation, but not at the station during the preceding hour, the fog or ice fog extending to a level above that of the observer
	41	Fog or ice fog in patches
	42	Fog or ice fog, sky visible, has become thinner during the preceding hour
	43	Fog or ice fog, sky obscured, has become thinner during the preceding hour
	44	Fog or ice fog, sky visible, no appreciable change during the preceding hour
	45	Fog or ice fog, sky obscured, no appreciable change during the preceding hour
	46	Fog or ice fog, sky visible, has begun or has become thicker during the preceding hour
	47	Fog or ice fog, sky obscured, has begun or has become thicker during the preceding hour
	48	Fog or ice fog, sky visible
	49	Fog or ice fog, sky obscured
		WW Code 50–59: Drizzle
	50	Drizzle, not freezing, intermittent – slight at time of observation
	51	Drizzle, not freezing, continuous – slight at time of observation
	52	Drizzle, not freezing, intermittent – moderate at time of observation
	53	Drizzle, not freezing, continuous – moderate at time of observation
	54	Drizzle, not freezing, intermittent – heavy at time of observation
	55	Drizzle, not freezing, continuous – heavy at time of observation
	56	Drizzle, freezing, slight
	57	Drizzle, freezing, moderate or heavy
	58	Drizzle and rain, slight
	59	Drizzle and rain, moderate or heavy
		WW Code 60–69: Rain
	60	Rain, not freezing, intermittent – slight at time of observation
	61	Rain, not freezing, continuous – slight at time of observation

Symbol	Code	Description
	62	Rain, not freezing, intermittent – moderate at time of observation
	63	Rain, not freezing, continuous – moderate at time of observation
	64	Rain, not freezing, intermittent – heavy at time of observation
	65	Rain, not freezing, continuous – heavy at time of observation
	66	Rain, freezing, slight
	67	Rain, freezing, moderate or heavy
	68	Rain or drizzle and snow, slight
	69	Rain or drizzle and snow, moderate or heavy

WW Code 70–79: Solid precipitation not in showers

Symbol	Code	Description
	70	Intermittent fall of snowflakes – slight at time of observation
	71	Continuous fall of snowflakes – slight at time of observation
	72	Intermittent fall of snowflakes – moderate at time of observation
	73	Continuous fall of snowflakes – moderate at time of observation
	74	Intermittent fall of snowflakes – heavy at time of observation
	75	Continuous fall of snowflakes – heavy at time of observation
	76	Diamond dust (with or without fog)
	77	Snow grains (with or without fog)
	78	Isolated star-like snow crystals (with or without fog)
	79	Ice pellets

WW Code 80–89: Showery precipitation; or precipitation with current or recent thunderstorm

Symbol	Code	Description
	80	Rain shower(s), slight
	81	Rain shower(s), moderate or heavy
	82	Rain shower(s), violent
	83	Shower(s) of rain and snow mixed, slight

Symbol	Code	Description
	84	Shower(s) of rain and snow mixed, moderate or heavy
	85	Snow shower(s), slight
	86	Snow shower(s), moderate or heavy
	87	Shower(s) of snow pellets or small hail, with or without rain, or rain and snow mixed, slight
	88	Shower(s) of snow pellets or small hail, with or without rain, or rain and snow mixed, moderate or heavy
	89	Shower(s) of hail, with or without rain or rain and snow mixed, not associated with thunder, slight

WW Code 90–94: Thunderstorm during the preceding hour, but not at the time of observation

Symbol	Code	Description
	90	Shower(s) of hail, with or without rain or rain and snow mixed, not associated with thunder, moderate or heavy
	91	Slight rain at the time of observation
	92	Moderate or heavy rain at the time of observation
	93	Slight snow, or rain and snow mixed, or hail, at the time of observation
	94	Moderate or heavy snow, or rain and snow mixed, or hail, at the time of observation

WW Code 95–99: Thunderstorm at the time of observation

Symbol	Code	Description
	95	Thunderstorm, slight or moderate, without hail but with rain and/or snow at the time of observation
	96	Thunderstorm, slight or moderate, with hail at the time of observation
	97	Thunderstorm, heavy, without hail but with rain and/or snow at the time of observation
	98	Thunderstorm, combined with duststorm or sandstorm at the time of observation
	99	Thunderstorm, heavy, with hail at the time of observation

◄ *Present weather: Showers of rain in preceding hour, not at time of observation.*

▲ *Past weather since last report: showers, with more than half the sky covered by cloud.*

Past weather

Simplified versions of the present weather symbols are used to indicate the weather since the last observation. The symbols are plotted in red/yellow to the bottom right of the station symbol.

PAST WEATHER

○   Cloud cover ½ or less of the sky throughout the appropriate period

◑   Cloud cover ½ or less for part of the appropriate period and more than ½ sky for part of the period

●   Cloud cover more than ½ of the sky throughout the appropriate period

⇥   Duststorm, sandstorm or blowing snow – visibility less than 1,000 m (3,280 ft)

≡   Fog or thick haze – visibility less than 1,000 m (3,280 ft)

❜   Drizzle

●   Rain

✷   Snow or rain and snow mixed

▽   Shower(s)

↯   Thunder, with or without precipitation

OTHER OBSERVATIONS

Manned weather stations carry out many other observations, some on a daily and some on a monthly basis. Such measurements are not used for forecasting purposes – and are not given on station plots, of course – but may be required for climatological studies or for the verification of specific forecasts (of rainfall, for example). Such observations may include soil temperatures (at various depths), evaporation measurements, and precipitation and sunshine measurements. The latter are of most immediate interest, so these will be described in detail.

Precipitation

The principal form of precipitation that needs to be measured is rainfall and determining this accurately is actually one of the most difficult measurements to obtain. Rain gauges are normally cylindrical, with a sharp-edged upper aperture turned to a very precise diameter (usually 127 mm (5 in)), so the collecting area is precisely known. Below the opening, a funnel with a narrow outlet reduces the likeli-

145

hood of evaporation and channels the precipitation into a suitable storage vessel. Once a day, the water is poured into a measuring cylinder, and the value read to the nearest 0.1 mm (approx. 0.01 in).

As we have seen earlier, the term precipitation includes not only rain, but also drizzle, snow and hail (and other rarer forms of ice). When any frozen precipitation is present, this is melted, measured and reported in terms of the 'rainfall equivalent'. At locations where high snowfall is expected, larger gauges are often used.

Many stations are also equipped with recording rain-gauges, which provide a continuous record and are thus particularly valuable in providing information on the timing and amount of precipitation throughout the day. These gauges are normally of the 'tipping-bucket' type, in which the water is directed to one of two small containers on a rocking arrangement. When one container fills, the rocker tips, emptying the full container and bringing the empty one into use. Each movement of the rocker is recorded. On older instruments a paper graph was used, somewhat similar to those used in barographs, but more modern forms record the movements electronically. The water is normally collected in a suitable vessel so that the overall daily rainfall may be checked against the cumulative amount shown by the tipping mechanism.

A major problem with raingauges is that they are easily affected by water splashing from the ground. To ensure accuracy, the gauge's opening is normally placed at a specific height of 30.5 cm (12 in) above the surrounding surface, the nature of which is also closely controlled. Similarly, the gauge must be well away from any trees, shrubs or buildings that might introduce eddies and falsify the readings. At exposed sites, raingauges are often surrounded by a circular protective mound some distance away, with a vertical inner face. Again the diameter of the inner wall and the slope of the outside surface are closely controlled to avoid inaccurate readings.

▼ A standard copper raingauge, showing the very accurately turned brass rim, and the closely mown grass surrounding it.

Sunshine

The conventional method of recording sunshine has been by the use of a Campbell-Stokes sunshine recorder. This consists of a glass sphere, which focuses sunlight on to specially treated card. The sunlight burns a trace into the card, which is replaced once a day, and the overall length of the trace is then measured and converted into duration. Although the readings obtained with this recorder are sometimes difficult to interpret, this method has been used for many years and its limitations are well understood. Various electronic devices have been introduced in recent years and will doubtless come into much greater use in future. Their readings are currently adjusted to give Campbell-Stokes-equivalent values.

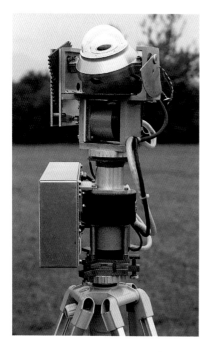

▲ *A Campbell-Stokes sunshine recorder. To allow for the changing elevation of the Sun throughout the year, three different types of sensitized card are used for specific seasons.*

◀ *An electronic solarimeter records both sunshine duration and intensity.*

A device that is also coming into increasing use is the electronic solarimeter. This determines the intensity of solar radiation, which is of particular interest in connection with horticulture and agriculture and also for climatological studies. Its records may, of course, be directly related to sunshine duration.

Amateur instruments

Amateurs who serve as voluntary observers, forwarding reports to their national meteorological office, require standard forms of equipment, observing screens, etc., such as those described earlier. Such stations are normally regularly inspected to ensure that the instruments and their exposure are satisfactory and that the reports are thus consistent with those from officially provided manned and automatic stations.

In recent years, however, a wide range of relatively inexpensive, electronic instruments has become available, many of which may be linked to a computer that will record observations at regular intervals. Records similar to those provided by barographs, thermographs and similar recording instruments are thus simply obtained. (An example of a pressure trace during the passage of a depression is shown on p.163.) Many such 'unofficial' observational reports are made available via the Internet.

Although the available electronic instruments are generally very accurate, the major problem consists of ensuring proper exposure. Unless, for example, the temperature sensors are mounted within a standard screen and that screen has an unrestricted airflow, temperature readings are likely to vary with wind direction and through radiation from nearby objects. So although readings of air temperature, humidity (dewpoint), maximum and minimum temperatures and pressure may be internally consistent, they cannot necessarily be compared with 'official' figures. The problems are particularly bad with respect to wind speed and direction, rainfall and sunshine. Few amateurs can install a 10-m (30-ft) mast for their anemometer and wind vane, for example, and they have to be content with mounting these devices somewhere on their roof, where the readings will be affected by turbulence and differ considerably depending on wind direction. Different, but similar, considerations apply to rainfall- and sunshine-recording devices. Such limitations must be borne in mind when considering long-term measurements or comparing records with nearby stations.

Hand-held devices are now commonly used for many outdoor activities, particularly for sailing, where hand-held anemometers have been used for some years. Many modern devices incorporate a propeller-type anemometer, which will give a useful indication of wind speed, provided the limitations of such equipment are borne in mind. Some hand-held instruments display a whole range of readings, including air temperature, dewpoint temperature, relative humidity, pressure, altitude, wind chill and wind speed.

▲ A typical form of hand-held meter, here shown displaying pressure and pressure tendency. Such devices not only incorporate an anemometer, but may also provide many other readings, including temperature, humidity, dew point, wind chill and altitude.

◀ *Mounting anemometers and wind vanes poses particular problems for amateurs. Even when the wind meets the anemometer before encountering the roof, the associated turbulence is likely to give erratic readings.*

SYNOPTIC CHARTS

The most informative charts of all are synoptic charts – charts that plot surface pressure at a specific observational time – particularly those that also show frontal systems. These are, of course, plotted from the pressure data reported by the various stations. It is perfectly possible to plot one's own synoptic charts from the information given in shipping forecasts, for example. Although not described in detail here, this is a particularly useful skill for sailors for occasions when it is not possible to download or otherwise receive officially prepared forecast charts, but when radio broadcasts are still available.

Charts showing the distribution of pressure at heights in the atmosphere other than the surface are known as isobaric charts. These are, of course, derived from data obtained from radiosonde ascents, from observations reported by aircraft and from satellite data.

Satellite images are now commonly shown in television weather forecasts, and are readily available from various sources on the Internet. While they often clearly show major systems such as depressions, some features that are clearly visible on synoptic charts do not reveal their presence in any way. It may be noted, incidentally, that black-and-white satellite images, i.e., those obtained in a single spectral channel, are often easier to interpret than the apparently more natural-looking, colour images obtained by combining several channels.

Regrettably, in recent years charts showing isobars have become less common in television weather forecasts, because broadcasters have come to believe that they are too complicated for people to understand. However, I would venture to suggest that if you should happen to see one, ignore what the television presenter is chattering about, and concentrate of fixing the pattern of isobars and fronts in your memory. It may not tell you everything about the way in which the weather will develop, but it provides a lot of significant information. Actual satellite images (or sequences of images) are also very informative.

We have seen much earlier (p.66) how closed isobars surround high- and low-pressure areas. These areas are usually marked 'H' and 'L', respectively, and the central pressure may also be shown. Depending on the weather service that has prepared the

MetMaps

For the area around the British Isles, base charts, known as MetMaps, have been prepared by the Royal Yachting Association and the Royal Meteorological Society. (URL = http://www.rmets.org/faq.php and click on the link) These charts show coastal outlines, named sea areas, and the location of coastal (i.e., reporting) stations. Because synoptic charts prepared using these blank forms have a fixed scale, it is possible to estimate approximate wind speeds, and the speed at which warm fronts or cold and occluded fronts are likely to move – all essential information for sailors, in particular. Conversely, a knowledge of the pressure gradient assists in the actual preparation of a synoptic chart.

charts, the interval between the isobars is typically 4 or 5 hPa (millibars). Occasionally, the interval may be as little as 2 hPA or as much as 8 hPA, but adjacent or alternate isobars are usually labelled to indicate the actual sea-level pressures.

High-pressure regions extend their influence as ridges, where the isobars are more strongly curved, and concave towards the centre of the high. Lows extend their influence as troughs, again where the isobars are more strongly curved and also concave towards the centre of the low. Troughs occur at both warm and cold fronts, but non-frontal troughs are also commonly found in the circulation around depressions. Particularly active non-frontal troughs are often indicated on isobaric charts that show frontal systems by a continuous black line. They are usually marked by greater wind speeds (closer isobars) and accompanied by increased cloudiness and precipitation. In a cold (usually polar) air mass, an active non-frontal trough may develop into a pronounced line of instability, or squall line, with vigorous cumulonimbus clouds or thunderstorms. Similarly, a trough-line extending away from the centre of a depression into the polar-air circulation on the polar and westward side of a low may sometimes develop into a small secondary depression (known as a polar-air depression), with its own circulation and pattern of precipitation (opposite). Such small low-pressure centres are distinct from the secondary depressions that develop on the main Polar Front (p.11).

▼ The principal features found on isobaric charts.

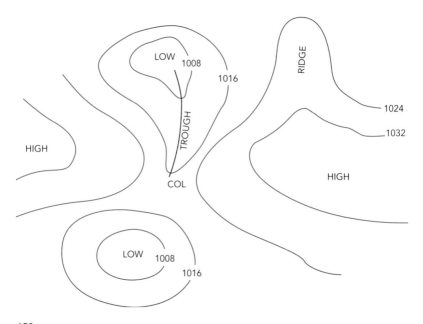

Another feature that is sometimes encountered on synoptic charts is a col. This is an area of slack pressure which lies between a pair of low-pressure areas and a pair of high-pressure areas. Although they are easy enough to see on a synoptic charts, cols do not show any distinct pattern of clouds on satellite images, in contrast to depressions, which are generally easy to locate, even in their early stages, when they may show just a 'comma-shaped' area of cloud. Because cols exist only when there is a temporary state of balance between the two pairs of high and low pressure centres, they are subject to sudden, dramatic changes in position – or may disappear completely – when the pressure pattern alters.

▲ *Two small polar-air depressions, with their own cloud patterns and closed circulations that have developed in the air flow around a relatively weak, primary depression (lower right).*

◄ An invisible col lies to the north-west of Scotland. One low is over the Bay of Biscay, and the other between Greenland and Iceland at top left. A ridge extends from the north-east towards northern Scotland and the other high pressure area is out over the Atlantic to the west.

Winds and wind speeds

As already described, the wind circulation around highs (anticyclones) is clockwise and that around lows (cyclones or depressions) anticlockwise – in the northern hemisphere and the opposite in the southern. The surface winds will be blowing at an angle to the isobars, in towards the centres of lows and out from the centres of highs. The closer the isobars, the stronger the wind. There is, however, a tendency for the winds around anticyclones to be somewhat stronger than one might expect from first glance at the spacing of the isobars. In cold air, the spacing usually gives a good indication of average wind strength, but the wind is likely to be very gusty. (This is because such air is usually unstable and the overturning motions bring higher air, with faster wind speeds, down towards the surface.) In warm air, by contrast, winds tend to be fairly steady and not as strong as might otherwise be expected.

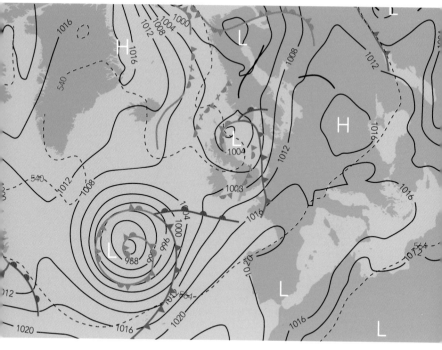

▲ *A typical synoptic chart, showing a deep, occluded low in mid-Atlantic. Another depression, with a developing occluded front over southern England, lies farther to the east. The significance of the dashed red contours is explained later (p.156).*

Depressions

Although high-pressure areas are readily visible on synoptic charts, because the weather associated with them is usually fairly quiet, not much may be inferred from the pressure pattern, apart from some indication of the wind speeds as just described. If a sequence of charts is available, this may give a clue as to how the high may move in the next few days. Otherwise, apart from the wind circulation around the high-pressure centre, practically all that may be inferred is the way in which any approaching depressions are likely to be deflected to the north or south of the high.

With depressions, however, far more information may be gained from synoptic charts. The significance of warm, cold and occluded fronts, and the accompanying changes in weather have already been described in detail (pp.66–75). Some official synoptic charts, such as those prepared by the United Kingdom's Meteorological Office, incorporate symbols to indicate whether fronts are developing or decaying. Although it is difficult to estimate the speed at which fronts are likely to move – unless the chart has been prepared to a fixed scale, as described earlier for MetMaps – as a general rule, cold and occluded fronts will move more rapidly than warm fronts. (This is why cold fronts catch up with warm

153

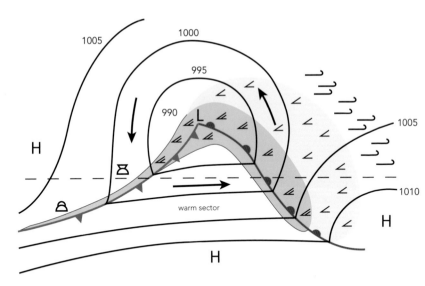

cirrus

thin altostratus

thick altostatus and nimbostratus

cumulonimbus

cumulus congestus

▼ *Typical precipitation patterns at a warm front (below right) and cold front (below left).*

fronts and lift the warm air away from the surface to form an occluded front.) The wind will always veer at a front (back in the southern hemisphere).

Important information about the likely motion of the low-pressure centre may be gained from the isobars in its warm sector. It will be recalled that the geostrophic wind – the wind away from the surface and unaffected by friction – flows essentially parallel to the isobars. Indeed, if the chart has been prepared to a known scale, it is possible to determine the geostrophic wind from the spacing of the isobars. As an approximate rule of thumb, it may be assumed that the centre of the depression is likely to move in a direction parallel to the isobars in the warm sector and at the same speed as the geostrophic wind. If a secondary depression (p.67) develops on the primary depression's trailing cold front, the two low-pressure centres will tend to rotate anticlockwise around one another (in the northern hemisphere), modifying the primary centre's anticipated motion.

Clouds and precipitation

The overall pattern of cloud cover and the likely precipitation may be inferred reasonably well from the generalized pattern found in most depressions, with alternating bands of heavier and lighter rain ahead of the surface warm

◄ *An idealized depression, showing a typical distribution of cloud types and areas of precipitation. The arrows show the geostrophic wind.*

▶ *The speckled appearance of the shower clouds in the maritime polar air is readily apparent, but the 'clear slot' that generally occurs immediately behind a cold front is not particularly well developed in this instance.*

▶ *Major cumulonimbus clusters and supercells are clearly visible in this image of the United States, with an enormous supercell system over parts of Texas, Oklahoma and Kansas.*

front (p.69) and light rain followed by a band of very heavy rain at a cold front. Although overall cloud cover may be inferred from satellite images, the exact type of cloud is usually far more difficult to assess, but the frequency and strength of the convective showers behind a cold front are often readily determined. Similarly, major cumulonimbus clusters – multicell storms and supercells (p.81) – will not be readily visible on synoptic charts, although any associated trough may be. Such features are, however, easily seen on satellite images.

Upper-air charts

As mentioned earlier, charts, known as isobaric charts, may be prepared for various altitudes in the atmosphere. One form plots the height of a particular pressure surface (such as 1,000 hPa, 500 hPa or 300 hPa) above mean sea level, and the 'thickness' – the difference between two such levels – usually with values given in decametres. (The dashed red contours on the chart on p.153, for example, show the thickness of the layer between 1,000 and 500 hPa.) This creates charts, similar to surface charts, that show troughs of low pressure, ridges, and pressure gradients. Such charts clearly show the location and strength of the jet streams and, more generally, the distribution of the major lobes of warm and cold air (p.18) along boundaries such as the Polar Front.

▼ *Height of 1,000 hPA pressure level (solid blue lines) and the 1,000–500 hPa thickness (dashed red lines).*

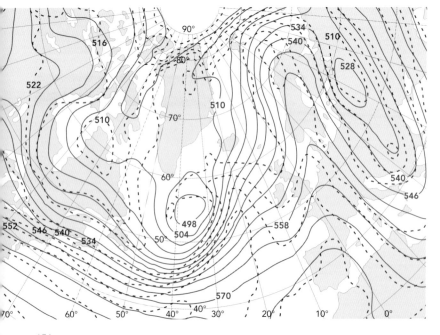

▶ *The cirrus clouds in the jet stream may be clearly seen overlying the warm front of this depression, and its path may be traced back along the cold front.*

Thickness contours are a direct reflection of the temperature of a layer of air. Because warm air expands, a region of warm air has a greater thickness than a corresponding region of cold air. In the chart shown opposite, for example, the solid lines indicate the height of the 1,000 hPa pressure level above mean sea level, and the dashed red lines the thickness between 1,000 and 500 hPa pressure levels. The coldest air lies over northern Labrador in Canada and over Finland, with a thickness of 510 dm (decametres). The warmest air is to the south, where 552 dm is the highest labelled contour. The difference between the warm air in the south and the cold air in the north has set up a considerable pressure gradient, with a band of strong winds snaking across the Atlantic and then up over Scandinavia. Over the central Atlantic this may be strong enough to be classed as a jet stream.

Because of the nature of their clouds (thin cirrus), jet streams are often difficult to detect on satellite images taken in a single spectral channel. Determination of cloud-top temperatures through the comparison of more than one channel usually provides evidence of their presence, however, as (naturally) do direct observations of wind speed at their level.

PROFESSIONAL FORECASTS

The flood of observational data from reporting stations is fed into the World Meteorological Organization's Global Telecommunication System, where the 'Main Trunk' route

▲ *This Meteosat image was obtained in a channel specifically designed to detect water vapour. Dark areas have lowest humidity. Such images are extremely useful in the preparation of weather forecasts and in tracking weather systems.*

runs from Washington, DC, in the USA, through Exeter in the United Kingdom, Offenbach in Germany and on to Melbourne in Australia. All of the world's main national and regional weather centres are linked to the primary route.

Some slight idea of the vast amount of data that is handled every day and used for specific forecasts may be gained from this approximate summary of observations handled by a typical national meteorological centre:

- over 40,000 reports from traditional land stations
- about 7,000 reports from moored and drifting oceanic buoys
- some 10,000 reports from shipping
- over 30,000 reports from aircraft
- about 1,200 temperature, humidity and pressure reports from radiosonde ascents
- around 700 reports of wind details from pilot-balloon ascents
- nearly 140,000 temperature and humidity determinations from polar-orbiting satellites
- about 8,500 determinations of wind strength and direction from geostationary satellites
- over 1,000,000 satellite observations that allow the determination of sea-surface winds
- nearly 600 determinations of wind speeds and direction from wind profilers operated across the USA.

The last source has not been described earlier, but consists of specialized radar installations that are designed to probe the atmosphere directly above the station. Measurement of the Doppler shift in the return signals enables the determination of wind direction and strength at different levels of the atmosphere. Some installations also include a microwave radiometer, which provides temperature and humidity profiles throughout the atmosphere.

All major meteorological centres now use numerical weather prediction (NWP) methods, whereby future states of the atmosphere are calculated by computer. Some of the world's fastest and most complex supercomputers are used for this purpose. The basic method consists of using a set of equations to calculate the changes that will occur in various parameters (such as pressure, temperature, humidity, wind speed and wind direction) over a given period of time in a model of the atmosphere. Computations are typically taken forward in 15-minute steps, and the calculations are continued to model the evolution of the atmosphere over the specific period of time required for a particular forecast. Such forecast intervals range from a few hours to a few days.

The synoptic reports are used to derive numerical values for the various elements at a set of grid points distributed around the world and at multiple levels within the atmosphere. The methods used by the European Centre for Medium-Range Weather Forecasting (ECMWF) in Reading in the United Kingdom may be taken as an example. (This is an international centre, supported directly by 18 European countries with co-operation agreements with another 10 European states.) ECMWF's overall global atmospheric model has a grid interval of 25 km (c.15.5 mi) and 62 levels. This is used to prepare daily, 3-day and 10-day forecasts. Forecasts out to as much as 21 days use a coarser grid with a 80-km (50-mi) mesh as do ensemble forecasts to 10 days. Global wave forecasts (out to 10 days) use a 55-km (34-mi) grid, and European-water forecasts (to 5 days) use the closer spacing of 27 km (17 mi).

The ensemble forecasts just mentioned are prepared using an interesting method inspired by chaos theory. It should be noted that the popular view of chaos theory is incorrect. It is often stated that small events can have large consequences, commonly called the 'Butterfly Effect' after the suggestion that the flap of a butterfly's wings in Brazil could unleash a tornado in Texas. In fact, a small effect could just as well prevent a major event. Chaos theory should, instead, be understood to imply that there are limitations on the accuracy of our observations of the atmosphere, on the equations used to compute its evolution and on the accuracy with which such computations may be carried out. The net result is that there must always be a degree of uncertainty in our predictions of the weather (or of any other similar system). Perhaps surprisingly, this may be turned to our advantage. In ensemble forecasting, a series of computa-

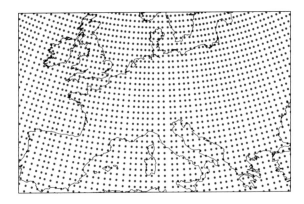

◀ An example of a relatively coarse grid with grid points at 50-km (31-mi) intervals over part of western Europe. Naturally, the spacing is not exact because the mesh has to cover an (approximately) spherical Earth.

tions is carried out, for the same projected period, but with slight variations in the initial input parameters. (The ECMWF method currently computes 51 projections.) If the simulations largely agree, then one has a high degree of confidence that the forecast is likely to be correct. If, however, they differ wildly, then it indicates that the state of the atmosphere is such that it may evolve in radically different ways. The method therefore provides a means of assessing which forecasts are most likely to be correct.

As increasingly powerful supercomputers become available, professional forecasters use them to improve the models used for weather and climatological forecasts most notably by reducing the interval between grid points and by introducing additional levels in the atmospheric models. At present, the grid spacings in use are too coarse to model many local features, such as hills, valleys, lakes, etc. that may have a considerable effect on local weather. Numerical weather prediction is therefore used to gain a broad outline of the development of the weather across a country (or in specific regions of a large country such as the USA). Individual human forecasters are then able to apply their knowledge of local conditions to provide more detailed, localized forecasts, which they are able to refine by reference to observations, such as radar plots of precipitation, that are available on a more or less 'real-time' basis. Such 'nowcasting' (p.125) is of immediate relevance to the general public.

MAKING A FORECAST

There are numerous factors that should really be taken into account in making a forecast, but this summary may be helpful. It also gives an indication of where the various items are discussed in greater detail. In general, try to look at the sky itself several times a day and check station plots and isobaric charts as frequently as you can.

The state of the sky and the clouds

Clouds give both a very important indication of the weather that exists at the time, and early clues to impending changes. This is particularly relevant to understanding when a depression is approaching. The various forms of cloud and optical phenomena are the first things that one should attempt to identify. Clouds are described on pp.27–53 and optical phenomena on pp.54–65. More general details of conditions found with various pressure systems are covered on pp.66–75.

The table given here can only offer an approximate guide to what may be expected. The major factors that will cause differences will be local ones, such as whether your location is to windward or downwind of high ground (pp.90–92), close to the sea (p.96) or subject to valley and mountain winds (pp.92–95).

CLOUDS

Clouds	Description	Forthcoming weather
cumulus humilis	scattered and moderate in size	fair weather, unless deep cumulus congestus or high cirrostratus appear
cumulus humilis and cirrus	cumulus fail to grow, cirrus increases	approaching warm front, especially if cirrus becomes cirrostratus; rain likely within 12 hours
cumulus congestus	towering cumulus	showers are possible within the next few hours
cumulus spreading into stratocumulus or altocumulus	clear sky decreases	precipitation unlikely, unless convection strong enough to break through layer; overcast likely to be slow to clear
cirrus	high thin clouds, few in number	generally fair, but weather may deteriorate if clouds increase and become organized
cirrus thickening to cirrostratus	halo may appear around Sun or Moon	warm front approaching with deteriorating weather; rain likely within 6–12 hours
cirrus or cirrostratus and altostratus	cloud increasing; strong upper wind spreads any contrails	warm front approaching, wind likely to increase, rain within 6–12 hours
jet stream cirrus	high, fast-moving bands of cirrus often with billows	a strong depression exists upwind and is approaching; cloud cover will increase; strong surface winds are likely in 10–15 hours
contrails	disappear rapidly	fine, major change unlikely within 24 hours unless strong solar heating produces major convective activity
contrails spreading	trails spread, become glaciated and persistent	increasing humidity at height; possible indication of approaching warm front with deteriorating weather

Clouds	Description	Forthcoming weather
altocumulus floccus or castellanus	clumps of altocumulus (often with virga) or lines of towers	thundery showers, possibly severe, likely within 24 hours
altocumulus and cirrocumulus	increasing in extent, becoming altostratus and cirrostratus	rapidly deteriorating weather, rain within 6–12 hours
altostratus and pannus	sky covered with altostratus, ragged pannus increasing	warm front nearby; rain imminent, intermittent at first, but becoming essentially continuous; wind likely to increase
nimbostratus	sky completely overcast	more-or-less continuous rain lasting for several hours
cumulonimbus	individual clouds of limited horizontal extent	light to moderate showers, gusts; active lifetime: 20–30 minutes
cumulonimbus	large clouds with several cells and/or anvils	heavy showers with possibility of hail and lightning; strong gusts; lifetime perhaps as long as a few hours
cumulonimbus	organized lines or clusters of massive clouds	extremely heavy showers, beginning with severe hail; if part of a cold front will be followed by scattered (and possibly heavy) showers
stratocumulus	low cloud with occasional breaks	no significant precipitation; likely to be slow to clear
mixed, stable clouds	stratocumulus, altostratus and cirrostratus patches	warm sector cloud; no significant change imminent and may persist for days
stratus	low cloud	no significant precipitation; if originated as fog, may clear later in day, otherwise persistent

Wind

The wind direction at various heights may be assessed (with care) from cloud movement and the relative motions at different heights are important in determining the likely weather, particularly around an approaching low-pressure system (pp.66–72). Station plots (pp.126–145) and synoptic charts (pp.149–157) are particularly useful here. Wind directions will alter with the approach and passage of frontal systems (p.70).

Pressure and pressure changes

The current pressure pattern is all important in assessing how it and the weather are likely to change. Interpreting synoptic charts (pp.149–157) is likely to be of greatest assistance here and station plots (pp.126–145) will add useful details – especially about local conditions – that will help to assess any localized variations. The pressure tendencies (p.136) will also give early clues as to the developments that are likely.

The approach of a depression is a major event (p.68) that may have drastic effects upon the weather over several days. It is heralded by falling pressure, while strong winds accompany steep rises and falls (p.152). A fall and subsequent steadying of pressure, or a fall and rise, will occur with the passage of a front. High pressure and slow changes are an indication of anticyclonic activity (p.75). Determining the

pressure in mbar

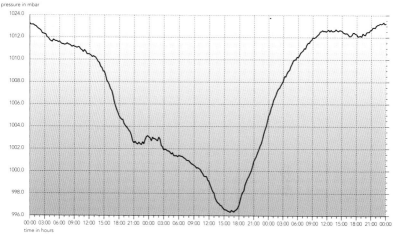

time in hours

▲ *The passage of a depression over three days from the pressure record. Pressure declines somewhat slowly as the warm front approaches, steadies slightly at the warm front (Day 1, 21:00) but continues to decrease until the cold front arrives, after which there is a rapid rise in pressure.*

direction of the geostrophic wind in the warm sector (p.154) will give a moderately good indication of the likely motion of the centre of a depression.

Air mass

The prevailing and encroaching air masses (p.16) will closely determine the pattern of weather, although this is frequently overridden by the influence of particular pressure systems, especially under anticyclonic conditions. The identification of an air mass may cause some slight difficulty, unless it is already known from other information, most notably temperature and humidity. For practical purposes it may be judged fairly well by the 'feel of the air'. The dry, extreme cold of continental polar (or even arctic) air is readily distinguished from the penetrating damp cold of maritime arctic air, while maritime polar air – although still damp – tends to have a more exhilarating feel. The humidity of maritime tropical air means that it may be easily told from the quality of the air.

Just going outside to judge the quality of the air and (if possible) comparing the impression gained with the officially reported air mass will soon enable you to estimate the likely air stream. Even the simple rule that in the northern hemisphere polar air will approach from directions between west through north to east (270°–360°/0°–90°) and tropical air between south-west through south to south-east (225°–180°–135°) will give an approximate idea of the origin. Obviously allowance must be made for regional and local effects. Isobaric charts – particularly a sequence – are of great assistance here in determining the previous path of an air mass and thus its characteristics.

Knowledge of the air mass is of particular use in forecasting likely maximum temperatures, described overleaf.

163

MAKING A FORECAST

Temperature

The development of daytime temperature will closely control cloud formation (p.23), which may lead to showers (p.76) or, with high temperatures, thunderstorms (p.79), hail (p.78), funnel clouds, such as waterspouts (p.85) or even tornadoes (p.84).

Maximum daytime temperature

Even if you have not compiled your own weather records, there is considerable practical value to having information on the maximum and minimum temperatures that occur in your own locality. Such figures for your own area will be readily available from your local library and you may well find suitable data on the Internet or in some of the books on the subject, although the latter are likely to contain data for just a few stations (perhaps only one) in your immediate area. Typical figures consist of averages for 10-day periods throughout the year. A typical extract from such a listing might be as shown in the table below (the figures are rounded to the nearest whole number).

Given such a table, you can easily see the average temperatures for any date. If you are able to obtain the maximum air temperature, either from your own observations, or from some other source, subtract the average temperature from the observed maximum air temperature. This gives a measure of the relative warmth, ranging from 'hot for the time of year' to 'cold for the time of year'. By considering the nature of the air mass, the sky conditions (form of cloud cover) and the strength of the wind, you can make an estimate of the likely temperature – again 'for the time of year'. This method is shown in graphical form opposite.

Overnight temperature

It is possible to forecast overnight temperatures, which is particularly useful for judging the likely occurrence of frost, by various methods but these involve knowing the dry- and wet-bulb temperatures in the early afternoon (at 15:00),

▶ *Determining likely maximum temperature. S = Some sunshine; C = Overcast or cloudy sky; L = Light winds only; W = Windy. Temperatures are amounts above or below normal for the time of year.*

MAXIMUM DAYTIME TEMPERATURE				
Period	**Maximum Temperature**		**Minimum Temperature**	
(approx. 10 days)	**°C**	**°F**	**°C**	**°F**
March 2–11	9	48	2	36
12–21	10	50	2	39
22–31	11	52	4	39
April 1–10	12	54	4	39
11–20	13	55	5	41
21–30	14	57	6	43
May 1–10	15	59	7	45

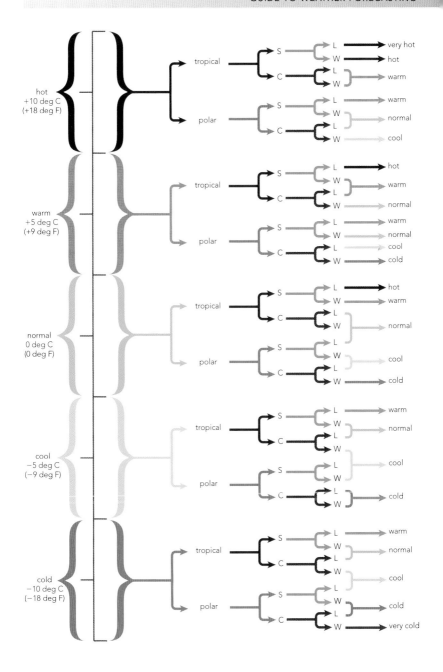

OVERNIGHT TEMPERATURE		
Northern hemisphere	**Interval between sunset and first measurement**	**Southern hemisphere**
December, January	1 hour	May, June
October, November, February	1.5 hours	March, April, July
March, April	2 hours	August, September

wind speed and in some methods, cloud cover and the number of days since rain last fell. If conditions are likely to remain clear overnight, however, a simplified method will give quite good results. To be able to use this method, it is necessary to know the times of sunset and dawn and to be able to take an accurate measurement of air temperature. After reaching a peak in early afternoon, the air temperature begins to decline and, a short time after sunset, adopts a constant rate of fall as the Earth's heat is radiated away to space. The method of determining the likely overnight minimum temperature therefore consists of taking two air temperature measurements an hour apart, with the first a certain time after sunset. In theory, the interval between sunset and the first measurement is governed by

▼ *Likely precipitation.*

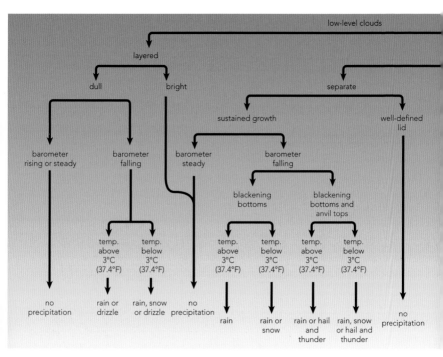

166

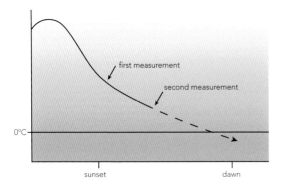

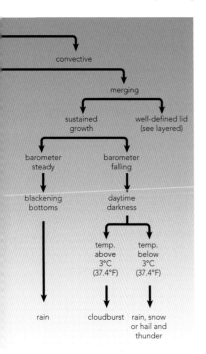

▲ *Estimating likely minimum overnight temperature. The first measurement is taken when the decline steadies about an hour after sunset.*

the site's latitude and the month. (The interval after sunset is less at lower latitudes.) However, the following method will work satisfactorily for most temperate latitudes.

When the rate of decline per hour becomes known from the second air temperature measurement, it is easy to see if the temperature will reach 0°C (32°F) during the night. In fact, the lowest temperature will be reached about one hour before dawn, after which the temperature will begin to recover, assuming, of course, that the sky remains fairly clear.

Likely precipitation

A graphical method of estimating likely precipitation is shown in the diagram on the left, and this is largely self-explanatory. It requires a knowledge of cloud types, pressure tendency and air temperature. It should be read in conjunction with the descriptions of individual cloud types, the details of the evolution of cloud cover in depressions and the development of showers and thunderstorms, all of which have been given earlier.

Humidity

High humidity, low temperatures and little wind indicate the possibility of fog, dew or frost (pp.99–105). High humidity and high temperatures, by contrast, are required for the formation of thunderstorms (p.79). The approach of a warm front is heralded by an increase in humidity, whereas low humidity generally accompanies anticyclonic conditions (p.75).

Local conditions

Great deviations from the conditions that prevail over an area may occur when local factors intervene. This applies, in particular, to cloud amounts (p.91), rainfall (p.92) and the formation of fogs and frosts (pp.99–105). Coastal areas may show considerable differences from weather inland and areas near hills or mountains may be greatly affected by valley, mountain or föhn winds (pp.92–97).

OBSERVING AND PHOTOGRAPHING THE SKY

Observing the sky, and particularly learning to recognize different cloud types, is very valuable for anyone with an interest in the weather. There are a few simple points that make it easier. It may seem to be stating the obvious, but you need to be able to see clouds and optical phenomena clearly. The greatest problem is often glare from the Sun (or occasionally the Moon). Hiding the Sun behind any object will usually make it much easier to make out faint iridescence or halo effects. Sunglasses – particularly the mirror type – are also useful in this respect. If possible, try to obtain small pieces of plastic polarizing material. One, used on its own, may be turned to emphasize the contrast between clouds and the sky. A pair of pieces may be turned relative to one another to provide a whole range of darkening, from almost complete transparency to fully blocking the light. Viewing reflections of clouds in a pool of water or against the dark glass used in many modern buildings will also frequently reveal details that are otherwise lost in glare.

It may sound slightly silly, but another useful tip is to examine clouds through binoculars. This will help you to see how clouds develop, and will show subtle changes that may otherwise be difficult to see. Examining the tops of cumulus congestus clouds, for example, will make it far easier to judge when they turn into cumulonimbus calvus, when their tops glaciate (i.e., freeze) and rain is likely. If you examine lenticular clouds you will probably be able to see from slight fluctuations in their extent how they are forming on the upwind side, and dissipating downwind – all the while remaining stationary in the sky. Do remember, however, that you must never look at the Sun through any optical instrument, so take great care if attempting to see features near the Sun itself. Never view the rising or setting Sun with binoculars, because invisible infrared radiation may be concentrated on the eye and cause damage before one is aware of it.

Photographing clouds is not particularly difficult, and is a wonderful way of becoming more familiar with cloud types. The most useful accessory is a polarizing filter, which will not only increase the contrast between clouds and the sky, which is often extremely useful in making subtle details visible but will also reveal detail in the clouds themselves. You will need to turn the filter in front of the lens to judge the best position for the effect you wish to achieve. If your camera does not have a filter mount, you can still hold a piece of polarizing material in front of the lens, although this is naturally slightly difficult to do by hand and you may need to hold the camera steady on a tripod.

For cloud details you may find that you need a telephoto lens of some sort, particularly for photographing certain

Measuring angles on the sky

It helps to be able to judge approximate angles on the sky, and this is significant in deciding whether clouds are strato-cumulus, altocumulus or cirrocumulus. A ruler, graduated in centimetres, held at arm's length may be used. Each one-centimetre graduation is almost exactly one degree. Even simpler, use your hand at arm's length:

1° = width of one finger-tip
7° = width over four knuckles
10° = width of clenched fist
22° = width of spread fingers
    (thumb to little finger)

The angular sizes of certain phenomena are summarized here:

52° = radius of secondary rainbow
46° = radius of outer (secondary) halo
42° = radius of primary rainbow
30° = altitude at which different cloud
    types are determined
22° = radius of inner (primary) halo
10° or more = width of stratocumulus
    cloudlets at 30° altitude
5–1° = width of altocumulus cloudlets
    at 30° altitude
1° or less = width of cirrocumulus
    cloudlets at 30° altitude
0.5° = diameter of Sun or Moon

1°

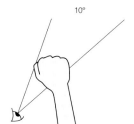

10°

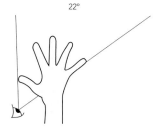

22°

optical phenomena. A wide-angle lens is often more useful, in that it will enable you to capture a whole expanse of sky. Panoramic images are also often very striking and relatively easy to obtain with digital cameras. Note, however, that the 'panoramic' setting on certain compact cameras does not give you any better image than the normal setting – it merely functions by masking the top and bottom of the image. The image size is exactly the same; is it simply that the 'panoramic' setting tells the printer to reproduce it in a 'letter-box' format. The same effect is achieved by enlarging part of a standard 35-mm frame.

FURTHER INFORMATION

Books

Brettle, M. & Smith, B., **Weather to Sail**, Crowood Press, Marlborough, 1999

Chaboud, R., **How Weather Works**, Thames & Hudson, London, 1996

Dunlop, S., **How to Identify Weather**, HarperCollins, London, 2002

Dunlop, S., **Oxford Dictionary of Weather**, 2nd edn, Oxford University Press, Oxford, 2008

Dunlop, S., **Wild Guide Weather**, HarperCollins, London, 2004

Eden, P., **Weatherwise**, Macmillan, London, 1995

Eden, P., **Weather Facts**, Oxford University Press, Oxford, 1996

Hamblyn, R., **The Cloud Book**, David & Charles, Newton Abbot, 2008

Hamblyn, R., **Extraordinary Clouds**, David & Charles, Newton Abbot, 2009

Harding, M., **Weather to Travel**, 3rd edn, Tomorrow's Guides, Hungerford, 2001

Henson, R., **The Rough Guide to Weather**, Rough Guides, London, 2002

Houghton, D., **Weather at Sea**, 2nd edn, Fernhurst Books, Arundel, 1998

Mayes, J. & Hughes, K., **Understanding Weather**, Arnold, London, 2004

Meteorological Office, **Cloud Types for Observers**, HMSO, London, 1982

Pearce, E.A. & Smith, C.G., **Hutchinson World Weather Guide**, 4th edn, Hutchinson, London, 1998

Pedgley, D., **Mountain Weather**, Cicerone Press, Milnthorpe, 1980

Reynolds, R., **Philip's Guide to Weather**, 2nd edn, Philip's, London, 2006

Watts, A., **Instant Weather Forecasting**, 2nd edn, Adlard Coles Nautical, London, 2000

Watts, A., **Instant Wind Forecasting**, 2nd edn, Adlard Coles Nautical, London, 2005

Williams, J., **The AMS Weather Book: The Ultimate Guide to America's Weather**, Univ. Chicago Press, Chicago, 2009

Journals

Weather, Royal Meteorological Society, Reading, UK (monthly)

Weatherwise, Heldref Publications, Washington D.C., USA (bimonthly)

Current weather

AccuWeather:
http://www.accuweather.com
  UK and Ireland:
  http://www.accuweather.com/ukie/index.asp?

BBC Weather:
http://www.bbc.co.uk/weather

CNN Weather:
http://www.cnn.com/WEATHER/index.html

Intellicast:
http://intellicast.com

ITV Weather
http://www.itv-weather.com

Unisys Weather:
http://weather.unisys.com

UK Meteorological Office:
http://www.metoffice.gov.uk

The Weather Channel:
http://www.weather.com/twc/homepage.twc

Wetterzentrale:
http://www.wetterzentrale.de/pics/Rgbsyn.gif

General information

HurricaneZone:
http://www.hurricanezone.net/

National Climate Data Center:
http://www.ncdc.noaa.gov
  Extremes: http://www.ncdc.noaa.gov/oa/
  climate/severeweather/extremes.html

National Hurricane Center:
http://www.nhc.noaa.gov/

Reading University:
http://www.met.reading.ac.uk/~brugge/
index.html

UK Weather Information Site:
http://www.weather.org.uk/

Unisys Hurricane Data:
http://weather.unisys.com/hurricane/
index.php

WorldClimate:
http://www.worldclimate.com

Meteorological Offices, Agencies and Organizations

Environment Canada:
http://www.msc-smc.ec.gc.ca/

European Centre for Medium-Range Weather Forecasting:
http://www.ecmwf.int

European Organization for the Exploitation of Meteorological Satellites:
http://www.eumetsat.int

Intergovernmental Panel on Climate Change: http://www.ipcc.ch

National Oceanic and Atmospheric Administration (NOAA):
http://www.noaa.gov/

National Weather Service (NWS):
http://www.nws.noaa.gov/

UK Meteorological Office:
http://www.metoffice.gov.uk

Analysis and Forecasts:
http://www.metoffice.gov.uk/public/
weather/surface-pressure/

Surface observations:
http://www.metoffice.gov.uk/public/
weather/observations/map

Weather observations:
http://www.metoffice.gov.uk/climate/
uk/wow.html

World Meteorological Organization:
http://www.wmo.ch

Satellite images

European Organization for the Exploitation of Meteorological Satellites:
http://www.eumetsat.int

Group for Earth Observation (GEO):
http://www.geo-web.org.uk

University of Dundee:
http://www.sat.dundee.ac.uk

University of Strasbourg
http://www-grtr.u-strasbg.fr

Societies

American Meteorological Society:
http://www.ametsoc.org/AMS

Australian Meteorological Society:
http://www.amos.org.au

Canadian Meteorological and Oceanographic Society:
http://www.meds.dfo.ca/cmos/

European Meteorological Society:
http://www.eumetsoc.org

Irish Meteorological Society:
http://www.irishmetsociety.org
   and http://www.iol.ie/~kcommins/MetSoc

National Weather Association, USA:
http://www.nwas.org/

New Zealand Meteorological Society:
http://metsoc.rsnz.org

Royal Meteorological Society:
http://www.rmets.org

MetMaps: http://www.rmets.org/faq.php,
   for link

TORRO: Tornado and Storm Research Organization: http://torro.org.uk

GLOSSARY

adiabatic Without the addition or loss of heat. Most parcels of air in the atmosphere rise and fall without exchanging heat with their surroundings.

aerosol Any minute solid or liquid particle that is suspended in the atmosphere.

air mass A body of air that has acquired specific characteristics, namely temperature and humidity, by remaining stationary over a particular area of the globe for some time. It tends to retain its temperature and humidity when it eventually moves away from the source area and largely determines the weather of any region it crosses.

anabatic Moving upwards. The term is typically applied to winds (such as a valley wind) or to the air at frontal systems.

anticyclone A region of high pressure where air has subsided from a higher level. Air flows out from the anticyclone over surrounding areas, with a clockwise circulation in the northern hemisphere.

anticyclonic Moving or curving in the same direction as air circulating around an anticyclone, i.e., clockwise in the northern hemisphere, anticlockwise in the southern.

antisolar point The point on the sky directly opposite to the position of the Sun, relative to the head of the observer.

backing An anticlockwise change in wind direction, e.g., from west, through south to east. The opposite of veering.

Beaufort scale A numerical scale for the description of wind speed, ranging from 0: calm; 1: 1–3 knots (0.3–1.5 m/s or about 1–3 mph) to 12: above 64 knots (above 33 m/s or c.73 mph).

Celsius The correct term for the temperature scale where the freezing and boiling points of water are 0°C and 100°C, respectively. (Frequently, and incorrectly, called 'Centigrade'.)

col An area with little air motion, located between a pair of low-pressure centres and a pair of high-pressure ones. Slight changes in the pressure pattern may cause a col to move rapidly or disappear.

continental climate A climate, typically encountered in continental interiors, where the characteristic features are extremely cold winters and hot summers. Such areas often experience low annual rainfall.

convection The transfer of heat by the motion within a fluid such as air or water. In the atmosphere convective motion is predominantly vertical. There are two forms of convection: 'natural convection' where parcels of air or thermals are free to move vertically driven by buoyancy effects; and 'forced convection' where the air is mixed mechanically by eddies.

Coriolis force The apparent force that deflects any moving object (such as the wind or an ocean current) away from a straight-line path. In the northern hemisphere it acts towards the right and in the southern, to the left. The faster the motion of the moving object, the greater the Coriolis force.

cyclone The technical term for a system in which air circulates around a low-pressure core. There are two distinct meanings: 1) a 'tropical cyclone', a self-sustaining tropical storm system also known as a hurricane or typhoon; 2) an 'extratropical cyclone' or depression, a low-pressure area, one of the major systems influencing the weather in temperate regions.

cyclonic Moving or curving in the same direction as air that flows around a cyclone, i.e., anticlockwise in the northern hemisphere, clockwise in the southern.

depression A commonly used term for a low-pressure area. Air flows into a depression and rises at its centre. The technical term for such a system is an 'extratropical cyclone'. The wind circulation around a depression is cyclonic (anticlockwise in the northern hemisphere).

dewpoint The temperature at which a volume of air, with a specific humidity, will reach saturation. At the dewpoint, water vapour begins to condense into droplets, giving rise to a cloud, mist or fog, or depositing dew on the ground.

föhn wind A hot, dry (and often desiccating) wind that descends on the leeward side of mountains. Having deposited most of its moisture on the windward slopes, it is much warmer and drier than at comparable levels on the opposite side of the range.

geostrophic A term applied to a (hypothetical) wind that flows parallel to the isobars on a chart. The wind at low-cloud height (600 m [2,000 ft]) corresponds approximately to a geostrophic wind.

hurricane A name used in the North Atlantic and eastern Pacific areas for a potentially destructive tropical cyclone.

instability The condition under which a parcel of air, if displaced upwards or downwards, tends to continue (or even accelerate) its motion. The opposite is stability.

inversion An atmospheric layer in which temperature remains constant or increases with height.

isobar A line that joins points on a weather chart that have the same barometric pressure.

jet stream A narrow band of high-speed winds that usually lies close to a break in the level of the tropopause. There are two main jet streams in each hemisphere (the Polar-Front and subtropical jet streams). Other jet streams exist in the tropics and at higher altitudes.

katabatic Moving downwards. Used primarily in connection with katabatic winds (fall winds) that sweep down from high ground and are normally initiated by low temperatures over the higher ground. The term is also applied to the motion of the air at certain frontal systems.

Kelvin A unit of heat (K), used to form a temperature scale beginning at absolute zero (−273.15°C −459.688°F). Temperatures are expressed in Kelvin units (e.g., 300 K), not as '300°K'.

lapse rate The rate at which temperature changes with increasing height. The lapse rate is positive when the temperature decreases with height and negative when the temperature increases.

latent heat The heat that is released when water vapour condenses or else freezes into ice crystals. It may be envisaged as the heat that was originally required for the process of evaporation or melting.

maritime climate A climate that is strongly influenced by a region's proximity to the ocean. Such climates generally have significant amounts of precipitation throughout the year. Winters are normally mild and the summers rarely experience extremely high temperatures.

mesosphere The atmospheric layer above the stratosphere, in which temperature decreases with height and reaches the atmospheric minimum at the mesopause, at an altitude of either 86 or 100 km (53 or 62 mi), depending on season and latitude.

mock sun A halo effect, consisting of a bright point of light, often slightly coloured and with a white tail. It lies at the same altitude as the Sun and approximately 22° away from it. Also known as a parhelion.

mountain wind A wind that blows down the length of a valley at night, primarily driven by the fact that the air over areas at a higher altitude cools more quickly than the air in the more protected valley.

occluded front A front in a depression system, where the warm air has been lifted away from the surface, having been undercut by cold air. The front may, however, remain a significant source of cloud and precipitation.

parhelion The technical term for the halo phenomenon otherwise known as a mock sun.

precipitation The technical term for water in any liquid or solid form that is deposited from the atmosphere and which falls to the ground. It excludes cloud droplets, mist, fog, dew, frost and rime, as well as virga.

pressure tendency The change in atmospheric pressure (which may be rising, falling or steady) during the previous three hours.

relative humidity The amount of moisture in the air, normally given as a percentage of the amount that the air would contain if fully saturated at a given temperature.

ridge An extension to an area of high pressure. On an isobaric chart it appears as a 'V'-shaped change in the direction of the isobars, pointing away from the pressure centre. Ridges are usually associated with increasingly fine weather.

stability The condition under which a parcel of air, if displaced upwards or downwards, tends to return to its original position rather than continuing its motion.

stratosphere The second major atmospheric layer from the ground, in which temperature initially remains constant, but then increases with

height. It lies between the troposphere and the mesosphere, with lower and upper boundaries of approximately 8–20 km (5–12.5 mi), depending on latitude, and 50 km (30 mi), respectively.

supercooling The conditions under which water may exist in a liquid state, despite being at a temperature below 0°C [32°F]. This occurs frequently in the atmosphere, often in the absence of suitable freezing nuclei. Supercooled water freezes spontaneously at a temperature of –40°C [–40°F].

synoptic chart A chart showing the values of a given property (such as temperature, pressure, humidity, etc.) prevailing at different observing sites at a specific time. An isobaric (pressure) chart is a typical example of a synoptic chart.

thermal A rising bubble of air, which has broken away from the heated surface of the ground. Depending on circumstances, a thermal may rise until it reaches the condensation level, at which its water vapour will condense into droplets, giving rise to a cloud.

tropopause The inversion that separates the troposphere from the overlying stratosphere. Its altitude varies from approximately 8 km (5.5 mi) at the poles to a maximum of 18–20 km (11.2–12.5 mi) over the equator.

troposphere The lowest region of the atmosphere in which most of the weather and clouds occur. Within it, there is an overall decline in temperature with height.

trough An extension of an area of low pressure, which results in a set of approximately 'V'-shaped isobars, pointing away from the centre of the low. A trough is often the location of increased cloudiness and rain.

valley wind A wind that blows up a valley during the day, driven by the greater heating of the air above the upper slopes, which draws air up from lower levels. Its counterpart is the night-time mountain wind.

veering A clockwise change in the wind direction, e.g., from east, through south to west. The opposite of backing.

virga Trails of precipitation (as ice crystals or raindrops) from clouds that do not reach the ground, melting and evaporating in the drier air between the cloud and the surface.

wind chill The loss of heat from the skin caused by the effects of wind. Even a moderate wind with create a heat loss that is as great as that occurring at a much lower temperature under calm conditions.

wind shear A change in wind direction or strength with a change of position. Vertical wind shear is defined as a change in wind strength with a change in height (usually an increase in speed with an increase in height). Horizontal wind shear is present if the wind strength alters from point to point at a fixed altitude.

zenith The point on the sky directly above the observer's head.

INDEX

ACKNOWLEDGEMENTS

Illustrations © Philip's, prepared
by Julian Baker, Stefan Chabluk
and Caroline Ohara.

CORBIS 30, 46t, 98; /Theo Allofs
42t; /Craig Aurness 38; /Joe Bator
77; /Tom Bean 31b, 62, 59b; /Niall
Benvie 52; /Dorothy Burrows 133;
/Michele Busselle 118; /DLILLC
46b; /Ric Ergenbright 47t; /Warren
Faidley 79; /First/Zefa 31t, 33t; /
Natalie Forbes 58; /Tom Fox 81;
/Lowell Georgia 78; /Philip Gould
38-9; /Michel Gounot 60;
/Eberhard Hummel 131/Image 100
36; /Jorma Jämsen 61; /Dewitt
Jones 63; /Layne Kennedy 48;
/Frans Lanting 101b; /Yang Lu 103;
/Medio Images 34; /Sally A
Morgan 85; /John Nakata 50b;
/NASA 26; /NASA HO/epa 7; /Eric
Nguyen 87t; /Philip Perry 59t;
/Bryan Pickering 49; /Pixland 64;
/Smiley N Pool 89; /Jim Reed 83,
107; /Reed Timmer/Jim Reed 87b;
/RH Productions 86; /Galen Rowell
37, 50t; /H David Seawell 111;
/Jonathan Smith 33b; /Richard
Hamilton Smith 35b, 43b, 44b, 45t,
102; /Scott Smith 104; /Dale C
Spartas 32; /Keren Su 51; /Wes
Thompson 42b /Craig Tuttle 55t;
/Randy M Ury100-101; /A&J
Verkalk 82/Visuals Unlimited 22,
47b; /Robert Weight/Eoscene 54;
/Staffan Widstrand 122; /Martin B
Withers 120

© Crown copyright 2007,
the Met Office Reproduced
with permission of the controller
of HMSO 121, 147

Storm R. Dunlop 19, 21, 27, 35t,
41, 43t, 44t, 55b, 56, 57, 68, 76,
124, 128, 148t

Eumetsat 12, 123, 158

JEL Hubert 130, 135, 146

Milan Konecny 155b

NASA 4, 26, 28

NOAA National Climate Data
Center 125

Peka Parviainen 53

Richard Paul Russell Limited 148b

MODIS Land Rapid Response
Team, NASA Goddard Space
Flight Center 109

University of Dundee 151, 152,
155t, 157